BURLEIGH DODDS SCIENCE: INSTANT INSIGHTS

NUMBER 131

Biological control of pests and diseases in horticultural crops

I0644820

burleigh dodds
SCIENCE PUBLISHING

Published by Burleigh Dodds Science Publishing Limited
82 High Street, Sawston, Cambridge CB22 3HJ, UK
www.bdspublishing.com

Burleigh Dodds Science Publishing, 1518 Walnut Street, Suite 900, Philadelphia, PA 19102-3406, USA

First published 2026 by Burleigh Dodds Science Publishing Limited
© Burleigh Dodds Science Publishing, 2026. All rights reserved.

This book contains information obtained from authentic and highly regarded sources. Reprinted material is quoted with permission and sources are indicated. Reasonable efforts have been made to publish reliable data and information but the authors and the publisher cannot assume responsibility for the validity of all materials. Neither the authors nor the publisher, nor anyone else associated with this publication shall be liable for any loss, damage or liability directly or indirectly caused or alleged to be caused by this book.

No part of this publication may be reproduced, stored in a retrieval system or transmitted in any form or by any means electronic, mechanical, photocopying, recording or otherwise without the prior written permission of the publisher.

The consent of Burleigh Dodds Science Publishing Limited does not extend to copying for general distribution, for promotion, for creating new works, or for resale. Specific permission must be obtained in writing from Burleigh Dodds Science Publishing Limited for such copying.

Permissions may be sought directly from Burleigh Dodds Science Publishing at the above address. Alternatively, please email: info@bdspublishing.com or telephone (+44) (0) 1223 839365.

Trademark notice: Product or corporate names may be trademarks or registered trademarks and are used only for identification and explanation, without intent to infringe.

Notice
No responsibility is assumed by the publisher for any injury and/or damage to persons or property as a matter of product liability, negligence or otherwise, or from any use or operation of any methods, products, instructions or ideas contained in the material herein.

British Library Cataloguing in Publication Data
A catalogue record for this book is available from the British Library

ISBN 978-1-83545-701-6 (Print)
ISBN 978-1-83545-702-3 (ePub)

DOI: 10.19103/9781835457023

Typeset by Deanta Global Publishing Services, Dublin, Ireland

Contents

© Burleigh Dodds Science Publishing Limited, 2026. All rights reserved.

© Burleigh Dodds Science Publishing Limited, 2026. All rights reserved.

Series list

Title	Series number
Sweetpotato	01
Fusarium in cereals	02
Vertical farming in horticulture	03
Nutraceuticals in fruit and vegetables	04
Climate change, insect pests and invasive species	05
Metabolic disorders in dairy cattle	06
Mastitis in dairy cattle	07
Heat stress in dairy cattle	08
African swine fever	09
Pesticide residues in agriculture	10
Fruit losses and waste	11
Improving crop nutrient use efficiency	12
Antibiotics in poultry production	13
Bone health in poultry	14
Feather-pecking in poultry	15
Environmental impact of livestock production	16
Sensor technologies in livestock monitoring	17
Improving piglet welfare	18
Crop biofortification	19
Crop rotations	20
Cover crops	21
Plant growth-promoting rhizobacteria	22
Arbuscular mycorrhizal fungi	23
Nematode pests in agriculture	24
Drought-resistant crops	25
Advances in detecting and forecasting crop pests and diseases	26
Mycotoxin detection and control	27
Mite pests in agriculture	28
Supporting cereal production in sub-Saharan Africa	29
Lameness in dairy cattle	30
Infertility and other reproductive disorders in dairy cattle	31
Alternatives to antibiotics in pig production	32
Integrated crop-livestock systems	33
Genetic modification of crops	34

© Burleigh Dodds Science Publishing Limited, 2026. All rights reserved.

© Burleigh Dodds Science Publishing Limited, 2026. All rights reserved.

© Burleigh Dodds Science Publishing Limited, 2026. All rights reserved.

© Burleigh Dodds Science Publishing Limited, 2026. All rights reserved.

Acknowledgements

Chapters in this Instant Insight are taken from the following sources:

Chapter 1 - Advances in biopesticides for insect control in horticulture
Chapter taken from: Collier, R. (ed.), Improving integrated pest management in horticulture, Burleigh Dodds Science Publishing, Cambridge, UK, 2022, (ISBN 978 1 78676 753 0; www.bdspublishing.com)

Chapter 2 - Advances in bioprotectants for plant disease control in horticulture
Chapter taken from: Collier, R. (ed.), Improving integrated pest management in horticulture, Burleigh Dodds Science Publishing, Cambridge, UK, 2022, (ISBN 978 1 78676 753 0; www.bdspublishing.com)

Chapter 3 Effective biocontrol of grape diseases: assessing and improving efficacy
Chapter taken from: Sharma, S. and Ansari, M. (eds.), Advances in bioprotection of plants against diseases, Burleigh Dodds Science Publishing, Cambridge, UK, 2025 (ISBN 978 1 80146 728 5; www.bdspublishing.com)

Chapter 4 - Use of biocontrol agents in fruit tree disease management
Chapter taken from: Xu, X. and Fountain, M. (eds.), Integrated management of diseases and insect pests of tree fruit, Burleigh Dodds Science Publishing, Cambridge, UK, 2019, (ISBN 978 1 78676 256 6; www.bdspublishing.com)

© Burleigh Dodds Science Publishing Limited, 2026. All rights reserved.

Chapter 1

Advances in biopesticides for insect control in horticulture

Travis R. Glare, Bio-Protection Research Centre, Lincoln University, New Zealand; and Aimee C. McKinnon, La Trobe University, Australia

1 Introduction

Horticulture, from the production through to the sale of fruit, vegetables and flowers, is an industry that is crucial for food production and human wellbeing. The science of horticulture is critical to the prosperity and health of people and the environment. With an increasing population, it is estimated that food production will need to increase by 70% by the middle of the century to meet demand (Davies, 2015). Opportunities exist to develop new high-value and high-nutrition horticultural produce on the decreasing area of land available for food production. However, horticulture is a more intensive farming approach than that used for broad-acre crops such as wheat and corn, which means that poor sustainability practices will have more severe consequences for future generations, although the area impacted is less than for some other crops. This includes approaches toward invertebrate pest control, as insects and other invertebrates cause significant damage and production losses across the horticulture sector around the world.

The application of synthetic pesticides has been the main approach to controlling pest insects in horticulture. However, applications of certain

http://dx.doi.org/10.19103/AS.2021.0095.01
© Burleigh Dodds Science Publishing Limited, 2022. All rights reserved.

pesticides on fruit crops, such as stone fruits, have pre-harvest intervals ranging between 7 and 28 days that preclude the use of treatments prior to fruit ripening, to minimise residues, during a period when the risk of pest outbreak is high (Hossain et al., 2006; Boston et al., 2020). Frequent use of broad-spectrum insecticides can also promote secondary pest outbreaks of insect pests such as mites, which are otherwise regulated by natural invertebrate predators (Epstein et al., 2000). Furthermore, the pest insect species of some horticultural crops occupy cryptic ecological niches. For example, *Carpophilus truncatus* beetles, pests in Australian almond orchards, live inside mummified nuts and in new nuts, so that insecticide sprays are often unable to reach them and are therefore ineffective (Boston et al., 2020). In the USA, pecan weevils are major pests that cause significant crop damage; the larvae and pupae occur inside the pecan nuts and in soil, so that control of the weevils using contact insecticides is restricted only to the adults (Mulder et al., 2012).

The development of insecticide resistance from the continual use of pesticides is also a key challenge. Development of new synthetic pesticides is now costly and difficult, partly due to the strict environmental and safety regulations involved when converting an active compound into a product (Qadri et al., 2020). For instance, a recent report estimated that in the USA, the development of a novel synthetic pesticide would cost in excess of $300 million and can take approximately 12 years (McDougall, 2016).

The withdrawal of pesticides, combined with health and environmental concerns and the rise in the resistance to current pesticides, is driving the need for new methods of control, including the growing use of biopesticides.

2 What are biopesticides?

Biopesticides have been loosely defined as biologically based pesticides (Glare, 2015). This can be quite a broad definition, including the incorporation of live organisms and their derivatives (e.g. metabolites), through to the use of semiochemicals in formulations. The use of plant extracts, for example, can fit into a broad definition of biopesticides, but authors often use the term to refer to a subgroup of biopesticides such as microbial-based biopesticides.

Kienwick (2007) limited the definition of biopesticides to only those based on microbes, while Copping and Menn (2000) included semiochemicals. The USA EPA defined biopesticides as 'certain types of pesticides derived from such natural materials as animals, plants, bacteria, and certain minerals' (http://www.epa.gov/pesticides/biopesticides). However, biopesticides are divided into three groups - microbial-based, plant-incorporated protectants (e.g. Bt corn) and biochemical - which encompass formulations of naturally occurring substances including plant extracts and pheromones.

© Burleigh Dodds Science Publishing Limited, 2022. All rights reserved.

Several novel approaches to using microbes for the control of pest insects have expanded the basic definition of biopesticides. Endophytes, microbes which can live inside plants without causing disease, can often confer resistance to pests and pathogens (McKinnon et al., 2017) and can be classed as a new biopesticide (Glare et al., 2012). Other novel biopesticide approaches include the use of RNAi, which is the use of the antiviral immune system of insects for post-transcriptional gene silencing (Vogel et al., 2019).

It is more common for researchers to define the term 'biopesticide' based on their own area of expertise and focus. In this chapter, we will focus on microbial-based biopesticides for use against pest insects. Paradoxically, 'microbial' in the area of entomopathogens generally includes nematode entomopathogens, which are not microorganisms *per se*, although, as the most commonly studied entomopathogenic nematodes that use symbiotic bacteria as an agent for lethality, this is understandable.

The term 'biopesticide' has also been questioned from a viewpoint of public understanding. The term 'pesticide' has become associated with chemical insecticides and the connotation of broad-spectrum killing. As such, it can be viewed negatively, even for biopesticides. The term 'biorationals' has been used (e.g. Rosell et al., 2008) to encompass biopesticides and other biologically based approaches.

The history of biopesticides is as long as the history of agriculture. The use of plant extracts is probably the earliest example of biopesticides, dating from about some 4000 years ago, with regard to extracts of the neem tree (El-Wakeil, 2013). The use of microbes to control insects is relatively recent; the first demonstration of a disease caused by a microbial infection in insects was for silkworms infected by *Beauveria bassiana*, published in 1835 by Agostino Bassi. It was not until the nineteenth century, after the discovery that microbes cause disease, that the idea of using microbes to control insects was proposed by Pasteur, with the first practical attempts in Russia in the 1890s (Zimmermann et al., 1995). The product Sporéine, based on the bacterium *Bacillus thuringiensis* (Bt), is thought to be the first microbial-based insecticidal biopesticide that was commercialised.

Since that time, biopesticide use has increased incrementally, especially for those based on Bt. However, in the last decade, with the reduction in available synthetic pesticides and more concern around sustainable plant production, the use of biopesticides and the availability of products have increased dramatically. There are several areas where biopesticide development has advanced significantly, which will be reviewed below.

3 Biopesticide use in horticulture

The global market for biopesticides has an estimated value of $3 billion USD and is expected to increase by 15% in the next few years, outpacing market

© Burleigh Dodds Science Publishing Limited, 2022. All rights reserved.

growth of novel synthetic pesticides by 10-fold (Damalas and Koutroubas, 2018). Although biopesticide sales have been increasing year on year, they still make up only a small percentage of the insecticide market worldwide. When looking at the current, most exploited, microbes used as biopesticides in horticulture, there is an indication of where some of the key challenges lie.

Microbial-based biopesticides for insect control have been based mainly on bacterial, nematode, viral and fungal entomopathogens. Although there have been products based on protozoa and microsporidia, the lack of rapid-kill and *in vitro* production methods has greatly limited their development. For viruses and nematodes, *in vitro* production limitations have restricted biopesticide development to only a few examples, such as nucleopolyhedroviruses (Nguyen et al., 2016) and the nematodes in *Heterorhabditis* and *Steinernema* which are associated with the entomopathogenic bacteria in the genera *Xenorhabdus* and *Photorhabdus* (Shapiro-Ilan et al., 2012).

Fungi in the genera *Beauveria* and *Metarhizium* are the most exploited entomopathogenic fungi that are currently formulated into biopesticides (Quesada-Moraga et al., 2020). Strains of these genera are the most common to infect insects naturally, with host ranges in the hundreds of species across a range of insect orders. These species are largely asexual, with just conidia produced in bulk on cadavers. They can also be grown easily on simple media, including grains (Grzywacz et al., 2014), which makes mass production possible. In general, strains of *Metarhizium* and *Beauveria* are safe for non-target organisms, such as predators and parasites, when applied in the field. This does not mean these other beneficial agents may not be infected in the laboratory in some cases, but these other organisms are far more resistant to infection, meaning they have a little impact in the field. The compatibility of these beneficial fungi with other environmentally sensitive control approaches has been exploited in integrated pest management (IPM) programmes (Glare et al., 2020b).

Insect-killing bacteria are known from diverse taxonomic groups, although only a few groups have been exploited as commercially available biopesticides (Jurat-Fuentes and Jackson, 2012), and even fewer used in horticulture. Spore-forming bacteria in the genera *Bacillus, Paenibacillus, Lysinibacillus* and *Brevibacillus* are the most commonly used for insect-control products, the majority of which target caterpillars or Diptera. Among the Proteobacteria, species of *Serratia, Pseudomonas, Chromobacterium* and *Burkholderia* have been used. Bt strains and subspecies have been the mainstay of biopesticides used in horticulture for over 70 years, with the first commercialised Bt sold in 1938, and still accounting for around 60–70% of all microbial-based biopesticide sales (Olson, 2015). Some bacteria act through infection, but in many cases, including Bt, most target mortality is due to toxins. The use of toxins, as opposed to microbes with an infectious mode of action, results in very different commercial

© Burleigh Dodds Science Publishing Limited, 2022. All rights reserved.

products. For example, in the case of Bt, in which there is no requirement for live cells in the final product. This greatly simplifies formulation and storage.

Entomopathogenic nematodes can be found in several major groups, but the most exploited for biopesticides are in the Heterorhabditidae and Steinernematidae families. These nematodes have symbiotic bacteria involved in the infection cycle. The two groups are not particularly closely related but do share common characteristics, since they are obligate parasites, horizontally transmitted, have an infective juvenile stage and are always associated with bacteria that are involved in infection (Koppenhőfer et al., 2020).

This group of nematodes has been the subject of commercial development, with 13 species now developed. Commercialisation has been aided by the often broad host range of the main nematode families and the development of *in vitro* mass-production methods (Shapiro-Ilan et al., 2012). Although the range of hosts can be broad in the laboratory, with over 200 species for *S. carpocapsae* across 10 insect orders (Koppenhőfer et al., 2020), for many species, the host range has not been investigated in the field. Infective juveniles are able to disperse actively and passively, which can greatly aid effectiveness.

Entomopathogenic viruses are known from 15 families, but most viruses used in biopesticides have been from the family Baculoviridae, containing the nucleopolyhedroviruses (NPV) and granuloviruses (GV), which target Lepidoptera (Srinivasan et al., 2019). These viruses have several attractive characteristics, principally that the infective virons are encased in a proteinaceous coat, ensuring more robust particles in the environment and suitability for formulation into biopesticides. The known viruses in these groups are host-specific, sometimes only known to infect a single lepidopteran species, meaning they are safe for non-target organisms. However, growing these viruses outside living cells is not currently possible, which greatly increases the cost of production and limits their use as biopesticides in many areas.

Some successful biopesticides have been developed based on baculoviruses. In Africa, commercial formulations of the *Maruca vitrata* Mavi NPV have been used in Benin, Niger and Nigeria (Srinivasan et al., 2019). NPV formulations have also been used against *Spodoptera*, *Plutella xylostella*, *Mamestra brassicae* and *Helicoverpa armigera,* including in citrus orchards in South Africa (Moore and Kirkman, 2010). In the Canary Islands, a *Chrysodeixis chalcites* NPV is under development as a commercial biopesticide for the pest of banana crops, *Chrysodeixis chalcites* (Bernal et al., 2018).

4 Key challenges in the successful use of biopesticides in integrated pest management programmes

There is a common list of challenges that have been reported to limit the uptake of microbial-based biopesticides. Some are based on the limitations of using

© Burleigh Dodds Science Publishing Limited, 2022. All rights reserved.

live organisms, such as lack of highly virulent agents, lack of environmental persistence, difficulties in economic mass production and application, short shelf life of formulations, lack of compatibility with other control options and limitations of the host range (either too broad or too narrow); but others are to do with regulation and registration of non-chemical pesticides in many regions (Table 1). One area that has been vexing for many proponents has been the reluctance of growers to use biopesticides. Also, this area has received less attention than the development of biopesticides, partly as it requires different skills from the biologists and technologists working with product development. These are the key challenges to overcome, and the specific issues for any biopesticide will be a mixture of these topics.

Furthermore, most bacteria-, virus-, nematode- and protozoa-based agents need to be ingested for action, as opposed to fungal entomopathogens and some nematode-transmitted bacteria. The need for ingestion of the toxin or cells places restrictions on the method of application and on formulation requirements. Agents used in biopesticides need to be easy to culture and have a shelf life suitable for the target pest.

Table 1 Common barriers to uptake of biopesticides

Barrier	Suggested solution
Lack of efficacy	Improve strain selection or use multiple agents in IPM
Short shelf life of biopesticides	Formulation improvements
Required to be easy to culture	In most cases, this is necessary for mass production to be economically viable
A common tendency to view biopesticides as chemical pesticide replacements - 'silver bullet' beliefs	More education for users with respect to benefits and limitations
Cryptic or soil-dwelling pests difficult to treat	Improvements in application methods informed by knowledge of pest ecology
Lack of environmental persistence of BCA	Formulation improvements
Lack of compatibility with other control options	Strain selection of BCA
Limitations of host range (either too broad or too narrow)	Improve and extend selection of agents
Difficulties in economic mass production	Selection of agents, or genetic modifications and/or research into production approaches
Regulation and registration barriers	Region-specific and can be addressed at that level
Slow to kill, so pest continues to damage plant	Strain selection, adjuvant/formulation to enhance mortality rate or new approaches such as mixed species of BCA or transgenic approaches
Reluctance of growers to use biopesticides	Field demonstration of efficacy

© Burleigh Dodds Science Publishing Limited, 2022. All rights reserved.

4.1 Agent selection

Selection of prospective entomopathogenic agents has traditionally relied on target invertebrate mortality rates observed under laboratory conditions (Waage, 1997). Frequently, efficacy is determined using a fundamentally reductionist approach, where both the agent and dose rate are tested in otherwise sterile and optimal conditions that favour the agent. This emphasis on mortality-by-dose arises from the type of biological control typically aimed for when using microbial-based biopesticides; that is, 'inundation' biological control (Eilenberg et al., 2001). Inundation biological control was defined by Eilenberg et al. (2001) as *'The use of living organisms to control pests when control is achieved exclusively by the released organisms themselves.'* In contrast with augmentation (or inoculation) biological control, which relies on growing and sustaining a viable population of the agent after release, the agent applied by inundation is not required to reproduce in order to achieve pest control (Eilenberg et al., 2001). The appealing aspect of this strategy is that the efficacy of the biological control agent (BCA) or its products is not dependent on the density of the target host population, as is the case with, for example, parasitoid wasps or other arthropod predators adopted as BCAs. There is, however, a common tendency to view BCAs in biopesticides as chemical pesticide replacements, offering an alternative but still providing a 'silver bullet' approach to the pest problem. If anything, this common perception has limited the development and use of otherwise effective agents in horticultural practices, because it is rare to achieve consistently high mortality rates in the field comparable to chemical pesticides (Chandler et al., 2011).

To improve field efficacy and use, novel biopesticides should be intentionally developed to contribute to an IPM 'toolkit,' where effective pest control relies on a combination of complementary or synergistic methods (Glare et al., 2020b). Part of the issue with the silver bullet mindset is that the full benefits and effectiveness of BCAs have not been incorporated in efficacy evaluations with focus on high virulence alone. For example, research indicates the need to consider ecological suitability, particularly with respect to biotic and abiotic interactions, during the initial process of agent selection (Hoy, 1992; Whipps, 1997; Deacon and Berry, 1993; Eilenberg et al., 2001; Meyling and Eilenberg, 2007; Bruck, 2010; Jaronski, 2010; Scheepmaker and Butt, 2010; Campos-Herrera et al., 2019; Martínez-Barrera et al., 2019; Glare et al., 2020b). Future biopesticides are therefore likely to be those developed for, and used in, a combination of biological control-type systems and/or will possess the capacity to confer multiple beneficial functions, such as biological control activity and plant growth promotion (Kowalska et al., 2020).

© Burleigh Dodds Science Publishing Limited, 2022. All rights reserved.

4.2 Bioassay

Typically, to assess the mortality rate of target pests by candidate biopesticide strains, laboratory bioassays are conducted. Bioassays are experiments that involve challenging the target host with varying dose concentrations of the candidate inoculum to determine the median lethal dose and mortality time after inoculation (Inglis et al., 2012; Thomas and Jenkins, 1997). Infective propagules of candidate entomopathogens are usually prepared in liquid suspensions and the concentrations calculated using histological techniques. To ensure consistent results, testing of potential candidates requires a method to deliver reproducible doses of the test substances (Spence et al., 2020). How the bioassay experiment is formally designed is also of critical importance, as the design can influence the consistency of results and therefore the success of the candidate(s) (Inglis et al., 2012). Bioassays require stringent experimental design with true replication because variables such as host species and age, host condition, size of bioassay container, incubation conditions and pathogen species/strain cultivation conditions can all affect results. Specific discussions of bioassays with entomopathogenic fungi can be found in Butt and Goettel (2000) and O'Callaghan et al. (2012) for experiments involving various entomopathogenic bacteria.

Even when robust laboratory bioassays are conducted, the results are often not reproducible when candidate agents are applied under field conditions to control the target pest (Roberts and St. Leger, 2004). This is because field conditions introduce more variables that can directly affect the host-pathogen interaction, the viability of the inoculum or the persistence of the pathogenic agent in the target environment. However, conventional dose-mortality bioassays may be useful in evaluating the effect of many different variables as treatments (Keyser, 2010). Using bioassays involving entomopathogenic fungi, studies have tested the effects of temperature on virulence (Fargues et al., 1997; Thomas and Jenkins, 1997; Brooks et al., 2004), the effect of repeated sub-culturing on virulence, and, if there is an effect, whether it can be improved by passing the pathogen through an insect host (Daoust and Roberts, 1982), the effect of insect thermal regulation on virulence (Blanford and Thomas, 2001; Ouedraogo et al., 2004) and the effects of propagule formulation (Prior et al., 1995; Thomas et al., 1995; Caudwell and Gatehouse, 1996).

4.3 Ecological suitability

While bioassays remain important to determine virulent entomopathogenic species or strains, simulating field conditions to further test virulence and ecological suitability in laboratory settings is challenging. Nevertheless, expanding screening criteria for entomopathogens to test certain factors

© Burleigh Dodds Science Publishing Limited, 2022. All rights reserved.

that are likely to be encountered under field conditions is crucial to increase the success of prospective biopesticides in horticultural systems (Onsongo et al., 2019; Blond et al., 2018). At a minimum, this may include heat tolerance, UV tolerance and vegetative growth-rate tests, which may be useful for shortlisting species or isolates to identify those more likely to be suitable in field conditions (Keyser, 2010). Primarily, it is important to understand the key factors that may affect the persistence of the agent once applied in the target environment. Where candidate isolates are required for application to soil, additional bioassays can be conducted using (non-sterile) soil taken from the target environment to study interactions that may occur with soil microbiota, soil structure or chemistry (Cory and Ericsson, 2010; Jackson et al., 2010; Jaronski, 2010). To date, there are few studies that have directly assessed the effect of the soil microbial community on inoculum or *vice versa* (Mayerhofer et al., 2017; Zitlalpopoca-Hernandez et al., 2017; McKinnon et al., 2018), yet the importance of plant-associated microbial communities for modulating plant health, including with respect to inducing insect-herbivore resistance, is currently highly topical (Dastogeer et al., 2020).

Conversely, plant species and plant rhizosphere systems can also directly influence the prevalence and persistence of entomopathogenic bacteria, fungi and nematodes. For example, plants have been shown to 'recruit' root-associated microbes following attack by insects or pathogens through the modulation of specific hormone-signalling pathways (Rudrappa et al., 2008; Lakshmanan et al., 2012; Lee et al., 2012). Rasmann et al. (2005) reported that an insect-induced plant root signal, (E)-β-caryophyllene, was emitted by maize roots in response to herbivory by beetle larvae *Diabrotica virgifera virgifera*. The (E)-β-caryophyllene was shown to strongly attract an entomopathogenic nematode. Some entomopathogenic fungi, particularly *Metarhizium anisopliae*, have been reported to be common in agricultural (tilled) soils compared to natural habitats, even though fungal diversity and abundance are normally higher in undisturbed soils (Bidochka et al., 2001; Meyling and Eilenberg, 2007; Inglis et al., 2008; Meyling et al., 2009).

Additionally, under field conditions, the presence of non-target insect hosts may affect the persistence of the biopesticide agent (Blond et al., 2018), as it can alter the concentration of the inoculum applied or its distribution in the environment. The presence of plants and plant microbiomes can also impact fungal persistence.

Aside from natural biotic and abiotic factors, biopesticides intended for use in horticulture need to be reasonably compatible with other components of the IPM toolkit that are used. For example, fungus-based biopesticides may be susceptible to certain fungicides used to control plant fungal pathogens, which can be problematic when these are used in the same IPM strategy (Chandler et al., 2011). Furthermore, IPM programmes that include combinations of

© Burleigh Dodds Science Publishing Limited, 2022. All rights reserved.

natural enemies to achieve pest control, such as different entomopathogens with invertebrate predators, must also consider and test for possible interactions between these biocontrol agents. Apart from testing prospective entomopathogens for pathogenicity to the invertebrate predators or parasitoids intended for use within the same IPM programme (Ekesi et al., 2007), certain combinations of natural enemies can result in 'non-additive effects,' where competition for the host as a resource reduces the overall efficacy of the pest management strategy – compared with when a single biocontrol agent is used (Ferguson and Stiling, 1996).

4.4 Production, formulation and application

Another essential aspect of agent selection for biopesticide development is the ease of production. Research has frequently involved and emphasised the optimisation of the mass production of inoculum, as well as agent stability and ease of delivery for application (Charnley and Collins, 2007). Accordingly, production systems have been devised to maximise inoculum yield at lower cost for use in inundative biocontrol applications (Wraight et al., 2001; Cliquet and Jackson, 2005; Jackson et al., 2010; Jaronski, 2010; Jaronski and Jackson, 2012). The underlying assumption has been that effective insect control may be achieved if copious amounts of inoculum can be feasibly produced to be applied at sufficiently high rates (Brownbridge et al., 2008; Jaronski, 2010). The variables encountered in the environment of application that may impact biocontrol activity have not been prioritised during product development, and thus in the formulation strategies (Jackson et al., 2010). However, recent research indicates that there is scope to modify selection and production to improve product quality (Lacey et al., 2015).

Nucelopolyhedroviruses and other agents that are obligate pathogens lead to particular problems around producing enough inoculum for use in biopesticides which are both economical and able to survive formulation and application. Large-scale production has proved elusive, with many efforts to grow pathogens *in vitro*, generally in insect cell lines. NPVs have been commercially produced based on infecting insects (e.g. Gychek, TM-Biocontrol (Bell et al., 1981; Reardon and Podgwaite, 1994; USDA, 2020)) or collecting infected cadavers from the field (Moscardi, 2007), but for many promising viral control agents, production is a limiting step (Bernal et al., 2018; Srinivasan et al., 2019). For the MaviMNPV for the control of *M. vitrata* in parts of Africa, community-based production is under development, where the mass production of the host insect and infection is completed at the village level (Srinivasan et al., 2019).

The current delivery methods are generally based on spray applications, similar to those used for synthetic pesticides or surface-applied granules for

© Burleigh Dodds Science Publishing Limited, 2022. All rights reserved.

the control of some soil-dwelling stages. These methods have been reviewed in detail (e.g. Bateman, 2003; Chapple et al., 2000). The main issue of any application process is getting sufficient viable inoculum to the target insect or site, and ensuring maximum persistence after application. In an IPM situation, it may include avoidance of application directly on other natural enemies or at certain times.

Attract-and-kill approaches have been developed for some insects, where a combination of semiochemical attraction to pathogen-laced traps is used (e.g. Yousef et al., 2018). The use of pathogen-impregnated traps, such as with entomopathogenic fungi, increases the subsequent chance of horizontal transfer to conspecifics and also has the advantages of using less of the pathogen than inundative approaches, with less residue in the crop (Quesada-Moraga et al., 2020).

The discovery and potential exploitation of any additional ecological roles of fungal and bacterial entomopathogens, as plant endophytes or rhizo- and phyllophane colonisers, is also under study as a mode of application (see Section 6.1).

4.5 Commercialisation: integration and safety

The most challenging aspects, that are to be considered early on, are the other control methods that are used for the target pest and whether the biopesticide is compatible (e.g. other biocontrol agents such as parasitoid wasps, lure traps, SIT, nematodes and fungi, compatibility with IPM-friendly pesticides). Is the biopesticide easy to use?

The compatibility of entomopathogens with other, especially biological agents, has been well documented (e.g. Glare et al., 2020b). Although all combinations need to be tested, generally parasitoids and predators are unaffected by field rates of biopesticides. For example, a recent study by Garriga et al. (2019) found that four predators and four parasitoids commonly used in horticulture were unaffected by *Heterorhabditis bacteriophora* or *Steinernema feltiae* treatments; and *Steinernema carpocapsae*, despite some indication of negative effects in the laboratory, did not impact the natural enemies in the field.

In most countries, biopesticides need to be registered under some form of pesticide regulations. For many years, biopesticides suffered under regulatory regimes which were established for chemical-based pesticides, which meant that some of the requirements for registration were inappropriate for biological-based products. This is slowly improving, with countries such as the USA and UK adopting specific registration requirements for biopesticides. However, a precautionary approach is often taken, which means unfamiliar products such as novel microbial-based biopesticides can face regulatory challenges.

© Burleigh Dodds Science Publishing Limited, 2022. All rights reserved.

Regulation can be tailored to biologicals, providing rigorous safe assessments, while not resulting in undue transactional costs (Grant and Gwynn, 2020). Regulators often have high field-efficacy targets of over 80% for pesticides, which, as discussed by Grant and Gwynn (2020), may not be appropriate for biopesticides. However, they can have lower efficacy, as long as the effect is to reduce a population below the economic damage threshold.

5 Integrated pest management

One of the trends in biopesticide use in horticulture is an increased focus on the integration of biopesticides into wider pest management strategies. The concept and application of IPM is not new, and there are good examples of commercially applied programmes that incorporate biopesticides. However, the use of multiple compatible pest control approaches into a single programme is increasing. Several recent reviews cover the potential and implementation of IPM with the inclusion of biopesticides (e.g. Glare et al., 2020b; Kiss et al., 2020). The compatibility of microbial pathogens with other organisms such as those providing ecosystem services (Srinivasan et al., 2019), including predators and parasites, has made IPM an attractive approach. There are various approaches that can be taken, from the rotational use of biopesticides with other softer control options, or of mixing agents together. As discussed by Kiss et al. (2020), implementation of IPM is easier in controlled situations such as glasshouses, than in open fields. IPM approaches can take advantage of the synergism that can occur between multiple agents. For example, neem has been shown to synergistically enhance NPV infections of *Spodoptera* and *Helicoverpa armigera* (Nathan and Kalaivani, 2006; Senthilkumar et al., 2008).

There are many examples of IPM using biopesticides in combinations. For example, Aristizabal et al. (2017) used the microbial biopesticides based on *Beauveria bassiana* GHA (BotaniGard) and *Metarhizium brunneum* F52 (Met-52 EC) in combinations with a horticultural oil (SuffOil-X) and azadirachtin (Molt-X) for the control of chilli thrips (*Scirtothrips dorsalis*) in Florida. In rotation, the agents gave as good a level of control as each alone, but tank-mixing the agents provided no additional benefit. Kiss et al. (2020) use the example of Ecophyto, a series of programmes by the French government to reduce the use of pesticides and plant protection products. One aim of the programmes is to shift from synthetic pesticides to biocontrol products. They quote a 33% reduction in the use of pesticides in horticultural production over 1900 test farms over 8 years.

Effective management of pest tephritid fruit fly species in various horticultural systems requires implementation of multiple techniques in an IPM program to reduce pest populations, minimise yield losses and enable growers to satisfy export market regulations (Aluja, 1996; Vargas et al., 2015;

© Burleigh Dodds Science Publishing Limited, 2022. All rights reserved.

Lezama-Gutierrez et al., 2000; Quesada-Moraga et al., 2006; Cossentine et al., 2010; Ekesi et al., 2011; Toledo et al., 2017; Yousef et al., 2017). Adult females cause damage by puncturing or 'stinging' ripening fruits, to lay eggs under the fruit skin. The eggs then hatch into larvae that feed on the flesh and these infested fruits subsequently rot (Aluja, 1996). Common techniques to achieve suppression of adult fruit flies and reduce produce losses include using food baits or attractants laced with insecticides. Food baits usually provide a protein source suitable for both female and male fruit flies, whereas pheromone attractants typically target male adult flies for a male annihilation technique (MAT) (Yousef et al., 2018). The MAT can also be used in combination with the sterile insect technique (SIT), where sterile male fruit flies are released to mate with females in order to limit damage from egg laying (Enkerlin et al., 2015). A biopesticide product, GF-120 (Dow AgroSciences, Indianapolis, IN), is an effective protein bait used for fruit flies which contains an insecticidal toxin, spinosad, that is derived from metabolites of the bacterium *Saccharopolyspora spinosa* (Sparks et al., 1995). In Hawaii, a GF-120 spinosad bait spray has been used effectively in combination with MAT and SIT, and with biological control (Mau et al., 2007). The efficacy of an application of GF-120 spinosad bait combined with an application of the entomopathogenic fungus *Metarhizium anisopliae* to soil was demonstrated by Ekesi et al. (2011) for the suppression of the tephritid fruit fly species, *Bactrocera dorsalis*, in Kenya. Prior studies also showed a reduction in adult emergence of different species of fruit flies in both laboratory and semi-field cage experiments after exposure to soil inoculated with *M. anisopliae* (Ekesi et al., 2007; Ouna, 2010).

Beauveria bassiana and *M. anisopliae* have also demonstrated pathogenicity against the Mediterranean fruit fly *Ceratitis capitata* adults, larvae, or pupae (Castillo et al. 1996; Quesada-Moraga et al., 2006; Soliman et al., 2020) and there have been studies conducted on the control of *Anastrepha fraterculus*, *A. ludens*, *B. oleae* and *B. tryoni* with different entomopathogenic fungi (Carswell et al., 1998; Lezama-Gutierrez et al., 2000;; Konstantopoulou and Mazomenos, 2005). The use of entomopathogenic nematodes against different tephritid species' larval stages has also been investigated (Langford et al., 2014; Soliman et al., 2014). Indeed, the application of these fungi or nematodes as biocontrol agents to soil can provide an additional control measure in IPM, because most control measures target adult flies and not larval stages.

5.1 Case study: Developing **Brevibacillus laterosporus** *for diamondback moth control in New Zealand*

Drawing on the authors' own experience, the development of the spore-forming bacterium, *Brevibacillus laterosporus* for diamondback moth (DBM) larval control highlights several of the factors that need to be considered in

© Burleigh Dodds Science Publishing Limited, 2022. All rights reserved.

biopesticide development. The bacterium is a well-known entomopathogen, although historic publications are limited (Ruiu, 2013). Among species in the genus, *B. laterosporus* is the only truly entomopathogenic species. Although strains can contain parasporal crystals and in some cases produce Cry toxins (Marche et al., 2018; Glare et al., 2020a), strains without parasporal crystals can also kill insects (Ruiu et al., 2007).

In New Zealand, screening was undertaken using a range of bacterial isolates recovered from *Brassica* plants, which included bioassays for activity against DBM larvae. The screening uncovered several *B. laterosporus* isolates that were highly toxic to larvae (van Zijll de Jong et al., 2016). Genome sequencing was conducted and several Cry-encoding toxin genes were found to be present (Glare et al., 2020a). However, these were later demonstrated not to be the main toxins for DBM (Ormskirk, 2017). As the species is also used as a probiotic for human consumption, safety was expected, and limited host range studies showed no activity against bees or other beneficial insects. *B. laterosporus* had also been previously isolated from the digestive tract of healthy foraging adult honey bees as a non-pathogenic endosymbiont in Europe (Gilliam and Valentine, 1976). Given the toxicity to a major insect pest such as DBM, and a mode of action that was not based on Cry toxin, the biological control activity differed sufficiently from that of Bt as a biological product. Additionally, the ease of mass production of the isolate enabled the initiation of a project to develop a novel biopesticide based on this New Zealand *B. laterosporus* strain. The New Zealand strains were determined to be an identifiable genotype compared to other *B. laterosporus* (Glare et al., 2020a). Much of the research for the development of the biopesticide is unpublished, but some can be found in the PhD thesis of Marsha Ormskirk (2017).

Following the standard requirements of a biopesticide, we investigated efficacy in bioassays, host range, production, formulation and field efficacy. Given the origin of the strains was the *Brassica* vegetable seed, we also investigated the endophytic potential of *B. laterosporus* in brassica. The bacteria were able to colonise brassica plants at a low level, and the presence of the bacteria endophytically had a small but significant effect on pupation of DBM (Ormskirk et al., 2019). The project team were investigating a range of potential BCAs for DBM control, so research was included that looked at the effect of combined agents in the laboratory and in the field. Combined applications with fungal and another bacterial entomopathogen demonstrated occasional synergistic effects in the laboratory. However, in the field, the effect was not significantly different from the activity observed of the single agents alone (Narciso et al., 2019).

The use of the novel New Zealand strains was patented, which gave a commercial partner confidence in taking the development to the pre-commercial stage. Field trials were conducted in New Zealand (Ormskirk et al.,

© Burleigh Dodds Science Publishing Limited, 2022. All rights reserved.

in preparation) and overseas, which indicated efficacy in a variety of situations. Currently, the company Ecolibrium Biologicals is registering a product in New Zealand, called Lateral, which is based on the bacteria. The development to the point of a product registration covers the classic steps in biopesticide development – efficacy against a target, safety, economic mass production, field efficacy and registration.

This case study demonstrates the general process and work involved to ensure the success of biopesticide product. However, candidate entomopathogenic agents, even when shown to be highly virulent to a target pest, do not often result in the development of a biopesticide product. Table 1 outlines some of the typical barriers that hinder the success of candidate biocontrol agents.

6 New approaches

6.1 The microbiome

Additional properties of some entomopathogenic fungi and their ability to colonise plants and contribute to resistance to plant diseases are leading to new considerations of the use of these agents. Many studies have shown how entomopathogenic fungi, in particular, and some bacteria can colonise plants, leading to a new consideration of control opportunities. This can even occur during inundative treatments, as spray application of some fungi for insect control led to transient endophytic colonisation that provided additional mortality to that caused by the direct spray (Garrido-Jurado et al., 2017; Resquin-Romero et al., 2016). Vega (2018) reviewed 85 papers covering 12 entomopathogenic fungal species in 6 genera, colonising 38 plant species in 19 families as endophytes. He concluded that colonisation was often localised and ephemeral, and a lack of understanding of the endophytism and plant responses, as well as the effect on insects, is holding back exploitation of this ecological niche.

The microbiomes of soils, plants and insects have all become topics of a lot of research focus, as techniques to directly evaluate components have become refined and cheaper. New approaches to biocontrol using microbes include consideration of the microbiome, whether to modify the insect microbiome for pest control or on/in plants (Lefort et al., 2017). For insects which depend on components of their microflora for deriving essential nutrients, knocking out specific microbes might act as a biocontrol approach. However, this requires the identification of the key components of the microbiome and then an understanding of how to specifically target certain microbes.

As more information accumulates, it is clear that interactions between microbes in the community are crucial to successful biocontrol, whether in

© Burleigh Dodds Science Publishing Limited, 2022. All rights reserved.

terms of interactions with an applied microbe, or regarding how the whole community might be manipulated for improved pest control outcomes. Microbes are known to provide nutrients to insects. Examples are *Buchnera aphidicola* in aphids and *Baumannia cicadellinicola* and *Sulcia muelleri* in the glassy-winged sharpshooter, *Homalodisca vitripennis* (reviewed in Gurung et al., 2019). Similar fungi can be involved in nutrient acquisition and host defence. One demonstrated benefit to plants of the *Metarhizium* infection of soil-dwelling insects was the transfer of nitrogen (Behie et al., 2012) and carbon (Behie et al., 2017) between the plant and the insect.

Viruses can sometimes manipulate host behaviour, in addition to being pathogenic. However, for many viruses, their interaction with insects is simply unknown. Similarly, protozoa and archaea are largely unexplored in this context (Gurung et al., 2019).

A review of patents in the area of beneficial endophytes (Ortega et al., 2020) listed 16 patents that mention 'pests,' including control of nematodes, although a few additional patents claim multiple activities of endophytes. The species included in the patents include well-published species with inverticidal properties (e.g. *Beauveria bassiana, Neotyphodium* sp., *Acremonium* spp., *Clonostachys rosea* and *Purpureocillium* sp.), but also groups better known as plant pathogens or saprophytes (*Aspergillus niger, Epichloë, Guignardia mangiferae, Muscodor* spp., *Sarocladium brachiariae, Cladosporium* sp., *Sphaerulina* sp., *Epicoccum* sp., *Stagonospora* sp., *Alternaria* sp., *Bipolaris* sp., *Stemphylium* sp., *Preussia* sp., *Curvularia* sp., *Penicillium* sp., *Phomopsis* sp., *Gibellulopsis* sp., *Fusarium* sp., *Chaetomium* sp. and *Cryptococcus* sp. (Ortega et al., 2020).

6.2 Genetic modification of microbes

Genetic modification of microbes has advanced greatly over the past decades and is seeing a step-change in possibilities with the use of CRISPR-Cas technologies. Modifying entomopathogens to overcome the current limitations to their use has been proposed and investigated for many years, although few genetically modified organisms have ever reached the field release and commercialisation stages. This is partly due to the current regulations in many countries.

Several studies have reported the expression of spider and scorpion toxin genes in *Metarhizium* and *Beauveria* that increase the virulence, even to the level of a single spore causing host death. Bilgo et al. (2017) reported that a range of toxins under a haemolymph-specific promoter, expressed in *Metarhizium pingshaense,* increased the virulence and speed-of-kill of mosquitoes, and a combination of two toxins, one from the North African desert scorpion *Androctonus australis (*AaIT) and another from the Australian Blue Mountains

© Burleigh Dodds Science Publishing Limited, 2022. All rights reserved.

funnel-web spider *Hadronyche versuta* (hexatoxin-Hv1a) acted synergistically. This combination was used in *M. pingshaense* in successful field trials against mosquitoes in Burkina Faso (Lovett et al., 2019). *Beauveria bassiana* was also modified to express scorpion toxin AaIT, for use against the adult mosquito *Aedes albopictus*. The lethal concentration required was 313-fold lower on day 4 and killed faster compared to the wild type (Deng et al., 2017). Similarly, *B. bassiana* expressing *AaIT* and *Pri*, a gene from *Metarhizium*, was more virulent to the Masson's pine caterpillar *Dendrolimus punctatus* and the wax moth *Galleria mellonella*, and killed 36–40% faster (Lu et al., 2008). Expressing a trypsin-modulating oostatic factor in *B. bassiana* increased virulence and reduced time-to-kill against *Anopheles gambiae* mosquitoes (Kamareddine et al., 2013).

Metarhizium anisopliae has been modified to overexpress a toxic protease, Pr1, which was coupled with a haemolymph-induced promoter. This overexpression led to enhanced virulence in the haemolymph of *Manduca sexta*, exhibited a 25% reduction in time-of-death, and reduced food consumption by 40%, compared to infections by the wild-type fungus (St. Leger et al., 1996). In *Metarhizium robertsii*, overexpression of the esterase gene *mest1* increased virulence against caterpillars, while expression of the same gene in *Metarhizium acridum*, where is it otherwise absent in the wild type, allowed infection of caterpillar hosts similar to *M. robertsii* (Wang et al., 2011b).

Increased persistence has also been the target of modification efforts. With *M. anisopliae*, melanin biosynthetic genes from *Alternaria alternata* were expressed, showing improvements in antistress capabilities and virulence against *Plutella xylostella* (Tseng et al., 2011).

Combining entomopathogenic components has also shown promise. Deng et al. (2019) reported that the expression of Bt Cyt2Ba toxin in *B. bassiana* showed increased virulence against *Aedes* mosquitoes. The lethal time-to-kill was reduced by up to 47%, although in other studies, expression of Bt toxins in NPVs did not improve the toxicity (Ribeiro and Crook, 1993; Martens et al., 1995).

Similarly, new approaches include combining entomopathogenicity with plant disease antagonists. Sorokan et al. (2020) reported that an endophytic strain of *Bacillus subtilis* that can inhibit mycelial growth of the potato pathogen *Phytophthora infestans* can also be modified to express the Cry1Ia insecticidal endotoxin from Bt. When tested, the modified strain showed insecticidal activity against the Colorado potato beetle *Leptinotarsa decemlineata*

The potential of modification has been demonstrated in areas such as increased virulence, reduced time-to-kill, altered host range, and in the case of some NPVs, reduced plant damage before insect death. Around 20 baculoviruses have been genetically modified to enhance performance (Scheepmaker et al., 2016). One area of focus in the 1990s was the viral *egt* gene, which blocks insect moulting to prolong caterpillar feeding. This

© Burleigh Dodds Science Publishing Limited, 2022. All rights reserved.

is a method for the virus to enhance the reproduction of viral particles in a fatter caterpillar but also results in more damage to the crop by infected caterpillars. Deletion in the *Autographa californica* nucleopolyhedrovirus (AcMNPV) of *egt* resulted in ~405× less feeding damage before insect death (Tomalski and Miller, 1992). This virus was also modified to express the insect toxin gene (Androctonus australis hector insect toxin [AaHIT]) from the scorpion *Androclonus australis,* which resulted in a 25% reduction in LT_{50} against *Trichoplusia ni*, compared to the wild-type virus, but with unaltered pathogenicity and host range. This combination was used in field trials in the 1990s, resulting in reduced crop damage and faster mortality of the caterpillars (Cory et al., 1994). Expression of an algal virus pyrimidine dimer-specific glycosylase cv-PDG in AcMNPV made the virus less susceptible to UV (3-fold decrease) and conferred a 16× increase in virulence (Petrika et al., 2003).

Toxins from straw itch mites (Tomalski and Miller, 1992), scorpions (Maeda et al., 1991; Elazar et al., 2001), sea anemone, and spiders have also been expressed in a range of NPVs, generally showing reductions in the time-to-kill by transgenic viruses compared to the wild type (reviewed in Kroemer et al., 2015). However not all gene deletions seem to affect the activity of NPVs (Bianchi et al., 2000; Kroemer et al., 2015).

Other modifications have included the deletion of a polyhedral envelope gene which resulted in a 6× increase in infectivity against *Trichoplusia* ni by AcMNPV (Ignoffo et al., 1995). The discovery of RNA interference as a gene-specific regulation mechanism has led to its use as a direct pest control approach (discussed below), but it has also been used to improve baculovirus efficacy (Kroemer et al., 2015).

The genes encoding Bt Cry toxins were identified many years ago, and some of these genes have been transgenically expressed in the most important crop plants in the world, such as maize, cotton and soybean. Bt-plants are the most common transgenic plants grown in the world. Similarly, expression of Cry toxin genes in other bacteria has been common, such as the expression of Bt cry1Ac in *Pseudomonas fluorescens,* with the recombinants highly toxic to *Helicoverpa armigera* (Ding et al., 2000). Recombinant *B. thuringiensis israelensis* with toxins from *B. sphaericus* were up to 5× more potent against *Culex* mosquitoes than wild-type strains (Park et al., 2003). The expression of genes encoding insect toxins in other beneficial bacteria has been explored. For example, a dipteran-active toxin gene from *B. thuringiensis israelensis* was expressed in the nitrogen-fixing bacterium *Bradyrhizobium* in nodules of pigeonpea, which provided protection against larvae of *Rivellia angulate* (Diptera) (Nambiar et al., 1990). Other potentially beneficial metabolites and enzymes can be expressed in insecticidal species for pest control enhancement. A Bt-expressing *Serratia marcescens* chitinase gene can help weaken the integrity of the insect gut

© Burleigh Dodds Science Publishing Limited, 2022. All rights reserved.

epithelium, giving better access for bacterial toxin genes to the insect, for stronger biocontrol activity (Okay et al., 2008).

The potential of genetic modification for the improvement of the active agents in biopesticides has been shown for decades. However, few commercially available products have been brought to market. Two Bt-based biopesticides were commercialised; CrymaxTM, based on a Bt strain with an additional Cry1C protein expressed, was commercialised by Certis; and LepinoxTM, another Bt which was modified to produce Cry 1Ac/1F proteins (Scheepmaker et al., 2016). The lack of GMO-based products is likely due to a combination of lack of public acceptance, restrictive regulations and insufficient benefits demonstrated to go through the regulatory process. There is also concern about non-target impacts (Azizoglu et al., 2020). With ever-increasing knowledge and development of novel techniques, and pressure for new control approaches, it is likely that GMO- based biopesticides will be sold in some countries in the near future.

6.3 The use of RNAi for insect control

A growing area of development for new biopesticide approaches is the use of RNAi for insect control. While not strictly a microbial-based approach, the approach can be used in genetically modified crops (expressing dsRNA) or applied as a formulation through a spray application, plant injection or irrigation (Yan et al., 2020). RNAi can also be transgenically expressed by microbes for enhanced effects. Expressing dsRNA in bacteria is considered the most efficacious production method (Joga et al., 2016). Sprayable RNAi has shown potential in pest management and has been trialled in horticulture. Topically applied RNAi can cause insect death, especially in insects such as aphids and mosquitoes, and has been demonstrated in greenhouse control of *L. decemlineata* (San Miguel et al., 2016). However, this approach has not been effective against all insects (Yan et al., 2020). In horticulture, application through irrigation has been used, such as for citrus tree protection (Hunter et al., 2012).

Delivery remains an issue, as formulation improvements are still needed to protect the RNAi from degradation. Nanoparticles have been investigated for this purpose, which can also promote translocation of the dsRNA across the peritrophic membrane or insect cuticle (Yan et al., 2020). There are still issues with RNAi as a pest management tool, including high production cost, sublethal effects against non-target species, resistance development and efficiency (Yan et al., 2020).

6.4 Production and formulation advances

Production is a major step in the development of biopesticides, when an organism or its bioactive metabolites are used inundatively. Scale-up is

© Burleigh Dodds Science Publishing Limited, 2022. All rights reserved.

expensive and so not all potential biopesticides are mass-produced, often because the market size is too small to justify the expense of production set-up. Growing massive amounts of cells can be challenging, and for some entomopathogens, *in vitro* growth is not possible currently (e.g. viruses, some nematodes and protozoa). Solid- and liquid-state fermentation are the basic methods to produce microbes for biopesticides, and in general, production has been incrementally improving. However, most of the technology has not changed much for decades.

For entomopathogenic fungi, local production has been a successful approach, especially where labour costs are low. Using production on solid substrates such as grains, fungi including *Beauveria* and *Metarhizium* can be produced cheaply in the locality of need. This approach has been widely used in the sugarcane-growing regions of Central and South America (Lacey et al., 2015), and there are other examples. However, in general, current mass solid fermentation methods for conidia production are not considered cost effective (Singhania et al., 2016).

For bacteria, liquid fermentation has been the most common approach, in the standard stirred batch fermentation. This has been particularly cost effective for the spore-forming bacilli, including Bt. For non-spore-forming bacteria, liquid fermentation has also been used, but post-production formulation and storage are more problematic without a spore stage (Johnson et al., 2001).

There are new production methods under investigation, but there must be a significant cost-saving incentive for companies to change processes, as capital equipment costs can be high. For example, approaches such as liquid interface production of fungi (e.g. Lopes et al., 2018) and production of microbes for biopesticides on waste materials have been investigated, as a method to reduce costs (e.g. Ballardo et al., 2016).

One area showing promise for the production of entomopathogenic fungi is the production of microsclerotia in liquid culture. Many fungi produce sclerotia - pigmented hyphal aggregates - which can survive environmental stresses better than most spores. It has been demonstrated that a number of entomopathogenic fungi can produce what are termed 'microsclerotia,' small sclerotia that can produce conidia after storage. The production of these microsclerotia was first demonstrated with *Metarhizium* spp. (Jackson and Jaronski, 2009; Song et al., 2014) and suggested as a useful new approach to producing propagules for biocontrol. Subsequent studies have also shown that *Beauveria* spp. can produce microsclerotia in submerged culture (Wang et al., 2011a; Villamizar et al., 2018). Huarte-Bonnet et al. (2019) have recently demonstrated that aggregation is dependent on FE2+ presence. Microsclerotia can be dried for formulation and storage, and then produce hyphae and sporulate. This makes microsclerotia a promising approach to liquid culture production of fungal biopesticides, which would be a cheaper approach than

© Burleigh Dodds Science Publishing Limited, 2022. All rights reserved.

the use of solid substrates for large-scale commercial production (Jackson and Jaronski, 2009; Song et al., 2014; Song, 2018).

7 Conclusion and future trends

Biopesticides are moving from niche products to more widespread use in large horticultural industries. Integrated pest management has been implemented in horticultural systems for many years, and in some cases, entomopathogens are highly effective tools in the arsenal of pest management systems. Integrated pest management systems are going to become much more common as synthetic pesticide use declines, and entomopathogens are well placed to be major components of new systems.

The increase in the use of IPM and of biopesticides is also due to improved technologies. The advent of new monitoring technologies, particularly automated systems, is benefiting the implementation of IPM, allowing more accurate prediction of pest peaks. In this review, we have outlined both a range of traditional approaches to IPM, such as combinations with other natural enemies and attractants, and examples of new technologies that have potential. Figure 1 provides a concept map that considers the various ideas for research and improvement of microbial-based biopesticides, which includes how candidate agents may fit within IPM programs.

In addition to the use of microbiome approaches, genetic engineering and RNAi, other new technologies are emerging. One method that has been used for a number of years is 'lure-and-kill' (El-Sayed et al., 2009), in which pheromone insect traps are laced with a microbial pathogen to kill

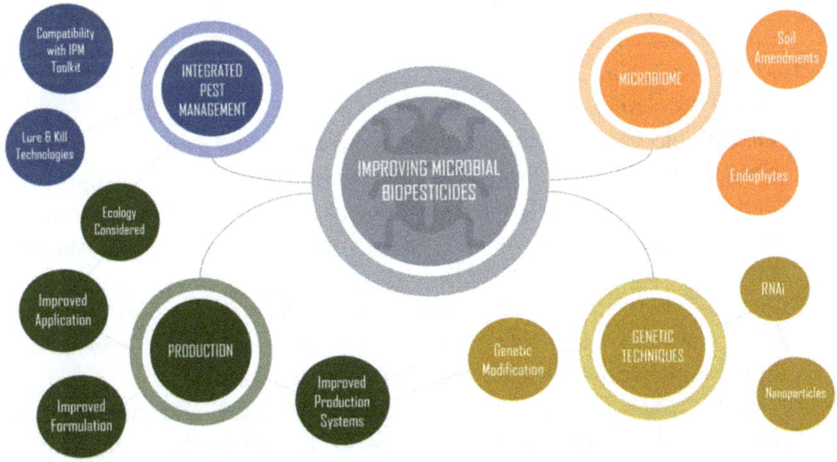

Figure 1 Concept map to consider how to improve biopesticide agent outcomes for insect pest management.

© Burleigh Dodds Science Publishing Limited, 2022. All rights reserved.

adults. A new approach by the American biotech company Pheronym is to use a novel nematode pheromone extract that increases the effectiveness of entomopathogenic nematodes, by increasing dispersal of the nematodes to hunt for insect targets. In a patent, the inventors claim that the use of the pheromone also expands the effective temperature range for the nematode pathogens. These Ascaroside pheromones stimulate dispersal of nematodes such as *Steinernema* spp., which both increase the dispersal of nematodes and reduce the quiescent period, which was reducing *S. carpocapsae* effectiveness at lower temperatures (Kaplan et al., 2020a). The team has also demonstrated the effect of the pheromone in microgravity (Kaplan et al., 2020b). This is an example of how a new understanding of the fundamentals of ecology and behaviour can be manipulated to improve biopesticide use in IPM.

Biopesticide use in horticulture will continue to increase, especially as the multinational pesticide companies move into these product opportunities and consumer demand drives the need for more environmentally friendly and safe pest control methods.

8 Where to look for further information

Further information on biopesticides can be found in many reviews and books devoted to the subject:

1 The Society for Invertebrate Pathology (http://www.sipweb.org/) is devoted to the study of entomopathogens.
2 The International Biocontrol manufacturers and the Research Institute of Organic Agriculture jointly run an Annual Biocontrol Industry meeting (https://www.abim.ch/about-abim.html).
3 International Organisation for Biological and Integrated Control: https://www.iobc-wprs.org/.
4 OECD biological pesticides website: http://www.oecd.org/chemicalsafety/pesticides-biocides/biological-pesticides.htm.
5 Information on regulations can be found on registration entities' websites, particularly the EPA USA (https://www.epa.gov/pesticides/biopesticides).
6 The FAO works to develop and promote IPM and support regulation of biocontrol products. Details can be found here: http://www.fao.org/agriculture/crops/thematic-sitemap/theme/pests/en/.
7 The EU database for plant protection products and guidance for biopesticides can be found here: http://ec.europa.eu/food/plant/pesticides/eu-pesticides-database/public/?event=homepage&language=EN.

© Burleigh Dodds Science Publishing Limited, 2022. All rights reserved.

8.1 Books

- Hall, F. R. and Menn, J. J. (Eds). (1999). *Biopesticides: Use and Delivery*. Springer.
- Birch, N. and Glare, T. R. (Eds). (2020). *Biopesticides for Sustainable Agriculture*. Burleigh Dodds Science Publishing Ltd., UK. 366 pp.

9 References

Aluja, M. (1996). Future trends in fruit fly management. In: McPheron, B. A. and Steck, G. J. (Eds), *Fruit Fly Pests: A World Assessment of Their Biology and Management*, St Lucie Press, Delray Beach, pp. 309–320.

Aristizabal, L. F., Chen, Y., Cherry, R. H., Cave, R. D. and Arthurs, S. P. (2017). Efficacy of biorational insecticides against chilli thrips, *Scirtothrips dorsalis* (Thysanoptera: Thripidae), infesting roses under nursery conditions. *J. Appl. Entomol.* 141(4), 274–284.

Azizoglu, U., Salehi Jouzani, G., Yilmaza, N., Baz, E. and Ozkok, D. (2020). Genetically modified entomopathogenic bacteria, recent developments, benefits and impacts: A review. *Sci. Total Environ.* 734, 139–169.

Ballardo, C., Abraham, J., Barrena, R., Artola, A., Gea, T. and Sánchez, A. (2016). Valorization of soy waste through SSF for the production of compost enriched with *Bacillus thuringiensis* with biopesticide properties. *J. Environ. Manage.* 169, 126–131.

Bateman, R. P. (2003). Rational pesticide use: Spatially and temporally targeted application of specific products. In: Wilson, M. (Ed.), *Optimising Pesticide Use*, John Wiley & Sons Ltd, Chichester, pp. 129–157.

Behie, S. W., Moreira, C. C., Sementchoukova, I., Barelli, L., Zelisko, P. M. and Bidochka, M. J. (2017). Carbon translocation from a plant to an insect-pathogenic endophytic fungus. *Nat. Commun.* 8, 14245.

Behie, S. W., Zelisko, P. M. and Bidochka, M. J. (2012). Endophytic insect-parasitic fungi translocate nitrogen directly from insects to plants. *Science.* 336(6088), 1576–1577.

Bell, R. A., Owens, C. D., Shapiro, M. and Tardif, J. R. (1981). Development of mass-rearing technology. In: Doane, C. C. and McManus, M. L. (Eds), *The Gypsy Moth: Research Towards Integrated Pest Management*. *For. Serv. Techn. Bull.* 1584, 599–632.

Bernal, A., Simón, O., Williams, T., Muñoz, D. and Caballero, P. (2018). Remarkably efficient production of a highly insecticidal *Chrysodeixis chalcites* nucleopolyhedrovirus (ChchNPV) isolate in its homologous host. *Pest Manage. Sci.* 74(7), 1586–1592.

Bianchi, F. J., Snoeijing, I., van der Werf, W., Mans, R. M., Smits, P. H. and Vlak, J. M. (2000). Biological activity of SeMNPV, AcMNPV, and three AcMNPV deletion mutants against *Spodoptera exigua* larvae (Lepidoptera: Noctuidae). *J. Invertebr. Pathol.* 75(1), 28–35.

Bidochka, M. J., Kamp, A. M., Lavender, T. M., Dekoning, J. and De Croos, J. N. A. (2001). Habitat association in two genetic groups of the insect-pathogenic fungus *Metarhizium anisopliae*: Uncovering cryptic species? *Appl. Environ. Microbiol.* 67(3), 1335–1342.

Bilgo, E., Lovett, B., Fang, W., Bende, N., King, G. F., Diabate, A. and St. Leger, R. J. (2017). Improved efficacy of an arthropod toxin expressing fungus against insecticide-resistant malaria-vector mosquitoes. *Sci. Rep.* 7(1), 3433.

© Burleigh Dodds Science Publishing Limited, 2022. All rights reserved.

Blanford, S. and Thomas, M. B. (2001). Adult survival, maturation, and reproduction of the desert locust *Schistocerca gregaria* infected with the fungus *Metarhizium anisopliae* var *acridum*. *J. Invertebr. Pathol.* 78(1), 1-8.

Blond, C. L. H., Ridgway, H. J., Brownbridge, M., Chapman, R. B., Condron, L. M., Saville, D. J. and Glare, T. R. (2018). Non-host larvae negatively impact persistence of the entomopathogen *Beauveria bassiana* in soil. *J. Invertebr. Pathol.* 156, 19-28.

Boston, W., Leemon, D. and Cunningham, J. P. (2020). Virulence screen of *Beauveria bassiana* isolates for Australian *Carpophilus* (Coleoptera: Nitidulidae) beetle biocontrol. *Agronomy.* 10(8), 1207.

Brooks, A. J., Muro, M. A., Burree, E., Moore, D., Taylor, M. A. and Wall, R. (2004). Growth and pathogenicity of isolates of the fungus *Metarhizium anisopliae* against the parasitic mite, *Psoroptes ovis*: Effects of temperature and formulation. *Pest Manage. Sci.* 60(10), 1043-1049.

Brownbridge, M., Townsend, R. J., O'Callaghan, M., Bell, N. L., Mander, C. and Jackson, T. A. (2008). Developing opportunities for entomopathogenic microbes and nematodes in crop protection. In: Butcher, M. R., Walker, J. T. S. and Zydenbos, S. M. (Eds), *Future Challenges in Crop Protection: Repositioning New Zealand's Primary Industries for the Future*, New Zealand Plant Protection Society Inc., Auckland, New Zealand, pp. 129-142.

Bruck, D. J. (2010). Fungal entomopathogens in the rhizosphere. *BioControl.* 55(1), 103-112.

Butt, T. M. and Goettel, M. S. (2000). Bioassays of entomogenous fungi. In: Navon, A. and Ascher, K. R. S. (Eds), *Bioassays of Entomopathogenic Microbes and Nematodes*, CAB International, Wallingford, pp. 141-195.

Campos-Herrera, R., Stuart, R. J., Pathak, E., El-Borai, F. E. and Duncan, L. W. (2019). Temporal patterns of entomopathogenic nematodes in Florida citrus orchards: Evidence of natural regulation by microorganisms and nematode competitors. *Soil Biol. Biochem.* 128, 193-204.

Carswell, I., Spooner-Hart, R. and Milner, R. J. (1998). Laboratory susceptibility of *Musca domestica* L. (Diptera: Muscidae) *and Bactrocera tryoni* (Frogatt) (Diptera: Tephritidae) to an isolate of *Metarhizium anisopliae* (Metsch.) Sorokin. *Aust. J. Entomol.* 37, 281-284. doi: 10.1111/j.1440-6055.1998.tb01584.x.

Castillo, M. A., Moya, P., Hernándeza, E. and Primo-Yúferab, E. (1996). Susceptibility of *Ceratitis capitata* Wiedemann (Diptera: Tephritidae) to entomopathogenic fungi and their extracts. *Biol. Control.* 19, 274-282.

Caudwell, R. W. and Gatehouse, A. G. (1996). Formulation of grasshopper and locust entomopathogens in baits using starch extrusion technology. *Crop Prot.* 15(1), 33-37.

Chandler, D., Bailey, A. S., Tatchell, G. M., Davidson, G., Greaves, J. and Grant, W. P. (2011). The development, regulation and use of biopesticides for integrated pest management. *Phil. Trans. Roy. Soc. Lond. B Biol. Sci.* 366(1573), 1987-1998.

Chapple, A. C., Downer, R. A. and Bateman, R. P. (2000). Theory and practice of microbial insecticide application. In: Lacey, L. and Kaya, H. (Eds), *Field Manual of Techniques in Invertebrate Pathology*, Kluwer Publishers, NLE, Dordrecht, pp. 5-37.

Charnley, A. K. and Collins, S. A. (2007). Entomopathogenic fungi and their role in pest control. In: Kubicek, C. P. and Druzhinina, I. S. (Eds), *Environmental and Microbial Relationships: The Mycota IV* (2nd edn.), Springer-Verlag, Berlin, pp. 159-187.

Cliquet, S. and Jackson, M. A. (2005). Impact of carbon and nitrogen nutrition on the quality, yield and composition of blastospores of the bioinsecticidal fungus *Paecilomyces fumosoroseus*. *J. Ind. Microbiol. Biotechnol.* 32(5), 204-210.

© Burleigh Dodds Science Publishing Limited, 2022. All rights reserved.

Copping, L. G. and Menn, J. J. (2000). Biopesticides: A review of their action, applications and efficacy. *Pest Manage. Sci.* 56(8), 651–676.

Cory, J. S. and Ericsson, J. D. (2010). Fungal entomopathogens in a tritrophic context. *BioControl.* 55(1), 75–88.

Cory, J. S., Hirst, M. L., Williams, T., Hails, R. S., Goulson, D., Green, B. M., Carty, T. M., Possee, R. D., Cayley, P. J. and Bishop, D. H. L. (1994). Field trial of a genetically improved baculovirus insecticide. *Nature.* 370(6485), 138–140.

Cossentine, J., Thistlewood, H., Goettel, M. and Jaronski, S. (2010). Susceptibility of preimaginal western cherry fruit fly, *Rhagoletis indifferens* (Diptera: Tephritidae) to *Beauveria bassiana* (Balsamo) Vuillemin Clavicipitaceae (Hypocreales). *J. Invertebr. Pathol.* 104(2), 105–109.

Damalas, C. A. and Koutroubas, S. D. (2018). Current status and recent developments in biopesticide use. *Agriculture.* 8(1), 13.

Daoust, R. A. and Roberts, D. W. (1982). Virulence of natural and insect-passaged strains of *Metarhizium-anisopliae* to mosquito larvae. *J. Invertebr. Pathol.* 40(1), 107–117.

Dastogeer, K. M. G., Tumpa, F. H., Sultana, A., Akter, M. A. and Chakraborty, A. (2020). Plant microbiome: An account of the factors that shape community composition and diversity. *Curr. Plant Biol.* 23, 100161.

Davies, F. T. (2015). Opportunities for horticulture to feed the world©. *Acta Hortic.* 1085, 455–458.

Deacon, J. W. and Berry, L. A. (1993). Biocontrol of soil-borne plant-pathogens: Concepts and their application. *Pestic. Sci.* 37(4), 417–426.

Deng, S. Q., Cai, Q. D., Deng, M. Z., Huang, Q. and Peng, H. J. (2017). Scorpion neurotoxin AaIT-expressing *Beauveria bassiana* enhances the virulence against *Aedes albopictus* mosquitoes. *AMB Exp.* 7(1), 121.

Deng, S. Q., Zou, W. H., Li, D. L., Chen, J. T., Huang, Q., Zhou, L. J., Tian, X. X., Chen, Y. J. and Peng, H. J. (2019). Expression of *Bacillus thuringiensis* toxin Cyt2Ba in the entomopathogenic fungus *Beauveria bassiana* increases its virulence towards *Aedes* mosquitoes. *PLoS Negl. Trop. Dis.* 13(7), e0007590.

Ding, Z., Zhang, J., Song, F., Huang, D. and Li, J. (2000). [Expression and synergism of two Cry insecticidal protein genes in *Pseudomonas fluorescens*]. *Wei Sheng Wu Xue Bao* 40(6), 573–578. [Chinese].

Eilenberg, J., Hajek, A. and Lomer, C. (2001). Suggestions for unifying the terminology in biological control. *BioControl.* 46(4), 387–400.

Ekesi, S., Dimbi, S. and Maniania, N. K. (2007). The role of entomopathogenic fungi in the integrated management of fruit flies (Diptera: Tephritidae) with emphasis on species occurring in Africa. In: Ekesi, S. and Maniania, N. K. (Eds), *Use of Entomopathogenic Fungi in Biological Pest Management*, Research Signpost, Trivandrum, India, pp. 239–374.

Ekesi, S., Maniania, N. K. and Mohamed, S. A. (2011). Efficacy of soil application of *Metarhizium anisopliae* and the use of GF-120 Spinosad bait spray for suppression of *Bactrocera invadens* (Diptera: Tephritidae) in mango orchards. *Biocon. Sci. Technol.* 21, 299–316.

Elazar, M., Levi, R.. and Zlotkin, E. (2001). Targeting of an expressed neurotoxin by its recombinant baculovirus. *J. Exp. Biol.* 204(15), 2637–2645.

El-Sayed, A. M., Suckling, D. M., Byers, J. A., Jang, E. B. and Wearing, C. H. (2009). Potential of "lure and kill" in long-term pest management and eradication of invasive species. *J. Econ. Entomol.* 102(3), 815–835.

© Burleigh Dodds Science Publishing Limited, 2022. All rights reserved.

El-Wakeil, N. E. (2013). Retracted article: Botanical pesticides and their mode of action. *Gesunde Pflanzen*. 65(4), 125–149.

Enkerlin, W., Gutiérrez-Ruelas, J. M., Villaseñor Cortes, A. V., Roldan, E. C., Midgarden, D., Lira, E., López, J. L. Z., Hendrichs, J., Liedo, P. and Arriaga, F. J. T. (2015). Area freedom in Mexico from Mediterranean fruit fly (Diptera: Tephritidae): A review of over 30 years of a successful containment program using an integrated area-wide SIT approach. *Fla Entomol.* 98(2), 665–681.

Epstein, D. L., Zack, R. S., Brunner, J. F., Gut, L. and Brown, J. J. (2000). Effects of broad-spectrum insecticides on epigeal arthropod biodiversity in Pacific Northwest apple orchards. *Environ. Entomol.* 29(2), 340–348.

Fargues, J., Ouedraogo, A., Goettel, M. S. and Lomer, C. J. (1997). Effects of temperature, humidity and inoculation method on susceptibility of *Schistocerca gregaria* to *Metarhizium flavoviride*. *Sci. Technol.* 7(3), 345–356.

Ferguson, K. I. and Stiling, P. (1996). Non-additive effects of multiple natural enemies on aphid populations. *Oecologia.* 108(2), 375–379.

Garrido-Jurado, I., Resquın-Romero, G., Amarilla, S. P., Rıos-Moreno, A., Carrasco, L. and Quesada-Moraga, E. (2017). Transient endophytic colonization of melon plants by entomopathogenic fungi after foliar application for the control of *Bemisia tabaci* Gennadius (Hemiptera: Aleyrodidae). *J. Pest Sci.* 90(1), 319–330.

Garriga, A., Morton, A., García-López, D. and García-del-Pino, F. (2019). Compatibility of entomopathogenic nematodes with natural enemies for horticultural pest control. *Biol. Control.* 138, 104050.

Gilliam, M. and Valentine, D. K. (1976). Bacteria isolated from the intestinal contents of foraging worker honey bees, *Apis mellifera*: The Genus *Bacillus*. *J. Invertebr. Pathol.* 28(2), 275–276.

Glare, T. R. (2015). Types of biopesticides. In: Nollet, L. and Rathore, H. S. (Eds), *Handbook of Biopesticides*, CRC Press, Boca Raton, pp. 7–24.

Glare, T. R., Caradus, J., Gelernter, W. D., Jackson, T., Keyhani, N., Köhl, J., Marrone, P., Morin, L. and Stewart, A. (2012). Have biopesticides come of age? *Trends Biotechnol.* 30(5), 250–258.

Glare, T. R., Durrant, A., Berry, C., Palma, L., Ormskirk, M. M. and Cox, M. P. (2020a). Phylogenetic determinants of toxin gene distribution in genomes of *Brevibacillus laterosporus*. *Genomics.* 112(1), 1042–1053.

Glare, T. R., McKinnon, A. C., Gwynn, R. L. and Brownbridge, M. (2020b). Advances in microbial control in integrated pest management: Entomopathogenic fungi. In: Kogan, M. and Higley, L. (Eds), *Integrated Management of Pest Insects: Current and Future Developments*, Burleigh Dodds Science Publishing, Cambridge, pp. 549-592.

Grant, W. and Gwynn, R. (2020). Improving regulatory approval process for biopesticides and other new biological technologies in agriculture. In: Birch, N. and Glare, T. R. (Eds), *Biopesticides for Sustainable Agriculture*, Burleigh Dodds Science Publishing, Cambridge, pp. 45–62.

Grzywacz, D., Moore, D. and Rabindra, R. J. (2014). Mass production of entomopathogens in less industrialized countries. In: Morales-Ramos, J. A., Guadalupe Rojas, M. and Shapiro-Ilan, D. I. (Eds), *Mass Production of Beneficial Organisms: Invertebrates and Entomopathogens*, Academic Press/Elsevier Inc., London, pp. 519–561.

Gurung, K., Wertheim, B. and Falcao Salles, J. (2019). The microbiome of pest insects: It is not just bacteria. *Entomol. Exp. Appl.* 167(3), 156–170.

© Burleigh Dodds Science Publishing Limited, 2022. All rights reserved.

Hossain, M. S., Williams, D. G., Mansfield, C., Bartelt, R. J., Callinan, L. and Il'ichev, A. L. (2006). An attract-and-kill system to control *Carpophilus* spp. in Australian stone fruit orchards. *Entomol. Exp. Appl.* 118(1), 11-19.

Hoy, M. A. (1992). Biological control of arthropods: Genetic engineering and environmental risks. *Biol. Control.* 2(2), 166-170.

Huarte-Bonnet, C., Paixão, F. R. S., Mascarin, G. M., Santana, M., Fernandes, ÉK. K. and Pedrini, N. (2019). The entomopathogenic fungus *Beauveria bassiana* produces microsclerotia-like pellets mediated by oxidative stress and peroxisome biogenesis. *Environ. Microbiol. Rep.* 11(4), 518-524.

Hunter, W. B., Glick, E., Paldi, N. and Bextine, B. R. (2012). Advances in RNA interference: dsRNA treatment in trees and grapevines for insect pest population suppression. Southwestern Entomol. 37, 85-87.

Ignoffo, C. M., Garcia, C., Zuidema, D. and Vlak, J. M. (1995). Relative in vivo activity and simulated sunlight-UV stability of inclusion bodies of a wild-type and an engineered polyhedral envelope-negative isolate of the nucleopolyhedrosis virus of *Autographa californica*. *J. Invertebr. Pathol.* 66(2), 212-213.

Inglis, G. D., Duke, G. M., Goettel, M. S. and Kabaluk, J. T. (2008). Genetic diversity of *Metarhizium anisopliae* var. *anisopliae* in southwestern British Columbia. *J. Invertebr. Pathol.* 98(1), 101-113.

Inglis, G. D., Enkerli, J. and Goettel, M. S. (2012). Laboratory techniques used for entomopathogenic fungi: Hypocreales. In: Lacey, L. A. (Ed.), *Manual of Techniques in Invertebrate Pathology* (2nd edn.), Academic Press, New York, pp. 189-253.

Jackson, M. A., Dunlap, C. A. and Jaronski, S. T. (2010). Ecological considerations in producing and formulating fungal entomopathogens for use in insect biocontrol. *BioControl.* 55(1), 129-145.

Jackson, M. A. and Jaronski, S. T. (2009). Production of microsclerotia of the fungal entomopathogen *Metarhizium anisopliae* and their potential for use as a biocontrol agent for soil-inhabiting insects. *Mycol. Res.* 113(8), 842-850.

Jaronski, S. T. (2010). Ecological factors in the inundative use of fungal entomopathogens. *BioControl.* 55(1), 159-185.

Jaronski, S. T. and Jackson, M. A. (2012). Mass production of entomopathogenic Hypocreales. In: Lacey, L. A. (Ed.), *Manual of Techniques in Invertebrate Pathology*, Academic Press, San Diego, pp. 257-286.

Joga, M. R., Zotti, M. J., Smagghe, G. and Christiaens, O. (2016). RNAi efficiency, systemic properties, and novel delivery methods for pest insect control: What we know so far. Front. Physiol. 7. doi: 10.3389/fphys.2016.00553.

Johnson, V. W., Pearson, J. F. and Jackson, T. A. (2001). Formulation of *Serratia entomophila* for biological control of grass grub. *N. Z. Plant Prot.* 54, 125-127.

Jurat-Fuentes, J. L. and Jackson, T. A. (2012). Bacterial entomopathogens. In: Vega, F. and Kaya, H. (Eds), *Insect Pathology* (2nd edn.), Elsevier, San Diego, CA, pp. 265-349.

Kamareddine, L., Fan, Y., Osta, M. A. and Keyhani, N. O. (2013). Expression of trypsin-modulating oostatic factor (TMOF) in an entomopathogenic fungus increases its virulence towards *Anopheles gambiae* and reduces fecundity in the target mosquito. *Parasit. Vectors.* 6, 22.

Kaplan, F., Perret-Gentil, A., Giurintano, J., Stevens, G., Erdogan, H., Schiller, K. C., Mirti, A., Sampson, E., Torres, C., Sun, J., Lewis, E. E. and Shapiro-Ilan, D. (2020a). Conspecific and heterospecific pheromones stimulate dispersal of entomopathogenic nematodes during quiescence. *Sci. Rep.* 10(1), 5738.

© Burleigh Dodds Science Publishing Limited, 2022. All rights reserved.

Kaplan, F., Shapiro-Ilan, D. and Schiller, K. C. (2020b). Dynamics of entomopathogenic nematode foraging and infectivity in microgravity. *NPJ Microgravity.* 6, 20.

Keyser, C. A. (2010). *Development of a Laboratory Based System for Selecting Insect Pathogenic Fungi with Greatest Potential for Success in the Field*, Vol. 604. All Graduate Theses and Dissertation. Utah State University, Utah.

Kienwick, S. (2007). Practicalities of developing and registering microbial biological control agents. *CAB Rev.* 2(13), 1–11.

Kiss, J., Istvan, S. and Delos, M. (2020). Implementing biopesticides as part of an integrated pest management (IPM) programme. In: Birch, N. and Glare, T. R. (Eds), *Biopesticides for Sustainable Agriculture*, Burleigh Dodds Science Publishing, Cambridge, pp. 21–44.

Konstantopoulou, M. and Mazomenos, B. E. (2005). Evaluation of Beauveria bassiana and B. brongniartii strains and four wild-type fungal species against adults of Bactrocera oleae and Ceratitis capitate. BioControl 50(2), 293–305. doi: 10.1007/s10526-004-0458-4.

Koppenhőfer, A. M., Shapiro-Illan, I. and Hiltpold, I. (2020). Advances in the use of entomopathogenic nematodes (EPNs) in suppressing crop pest insects. In: Birch, N. and Glare, T. R. (Eds), *Biopesticides for Sustainable Agriculture*, Burleigh Dodds Science Publishing, Cambridge, pp. 195–232.

Kowalska, J., Tyburski, J., Matysiak, K., Tylkowski, B. and Malusá, E. (2020). Field exploitation of multiple functions of beneficial microorganisms for plant nutrition and protection: Real possibility or just a hope? *Front. Microbiol.* 11, 1904–1904.

Kroemer, J. A., Bonning, B. C. and Harrison, R. L. (2015). Expression, delivery and function of insecticidal proteins expressed by recombinant baculoviruses. *Viruses.* 7(1), 422–455.

Lacey, L. A., Grzywacz, D., Shapiro-Ilan, D. I., Frutos, R., Brownbridge, M. and Goettel, M. S. (2015). Insect pathogens as biological control agents: Back to the future. J. Invertebr. Pathol. 132, 1–41. doi: 10.1016/j.jip.2015.07.009.

Lakshmanan, V., Kitto, S. L., Caplan, J. L., Hsueh, Y. H., Kearns, D. B., Wu, Y. S. and Bais, H. P. (2012). Microbe-associated molecular patterns-triggered root responses mediate beneficial rhizobacterial recruitment in *Arabidopsis*. *Plant Physiol.* 160(3), 1642–1661.

Langford, E. A., Nielsen, U. N., Johnson, S. N. and Riegler, M. (2014). Susceptibility of Queensland fruit fly, *Bactrocera tryoni* (Froggatt) (Diptera: Tephritidae), to entomopathogenic nematodes. *Biol. Control.* 69, 34–39.

Lee, B., Lee, S. and Ryu, C. M. (2012). Foliar aphid feeding recruits rhizosphere bacteria and primes plant immunity against pathogenic and non-pathogenic bacteria in pepper. *Ann. Bot.* 110(2), 281–290.

Lefort, M.-C., Boyer, S. and Glare, T. R. (2017). A response to Pennisi: "How do gut microbiomes help herbivores," a hint into next-generation biocontrol solutions. *Rethink. Ecol.* 1, 9–13.

Lezama-Gutierrez, R., Trujillo-de la Luz, A., Molina-Ochoa, J., Rebolledo-Dominguez, O., Pescador, A. R., López-Edwards, M. and Aluja, M. (2000). Virulence of *Metarhizium anisopliae* (Deuteromycotina: Hyphomycetes) on *Anastrepha ludens* (Diptera: Tephritidae): Laboratory and field trials. *J. Econ. Entomol.* 93(4), 1080–1084.

Lopes, R. B., Faria, M. and Glare, T. R. (2018). A novel biphasic fermentation system for the production of aerial conidia of entomopathogenic fungi utilizing surface tension. *Appl. Environ. Microbiol.* 126, 155–164.

© Burleigh Dodds Science Publishing Limited, 2022. All rights reserved.

Lovett, B., Bilgo, E., Millogo, S. A., Ouattarra, A. K., Sare, I., Gnambani, E. J., Dabire, R. K., Diabate, A. and St. Leger, R. J. (2019). Transgenic *Metarhizium* rapidly kills mosquitoes in a malaria-endemic region of Burkina Faso. *Science*. 364(6443), 894-897.

Lu, D., Pava-Ripoll, M., Li, Z. and Wang, C. (2008). Insecticidal evaluation of *Beauveria bassiana* engineered to express a scorpion neurotoxin and a cuticle-degrading protease. *Appl. Microbiol. Biotechnol.* 81(3), 515–522.

Maeda, S., Volrath, S. L., Hanzlik, T. N., Harper, S. A., Majima, K., Maddox, D. W., Hammock, B. D. and Fowler, E. (1991). Insecticidal effects of an insect-specific neurotoxin expressed by a recombinant baculovirus. *Virology*. 184(2), 777–780.

Marche, M. G., Camiolo, S., Porceddu, A. and Ruiu, L. (2018). Survey of Brevibacillus laterosporus insecticidal protein genes and virulence factors. *J. Invertebr. Pathol.* 155, 38–43.

Martens, J. W., Knoester, M., Weijts, F., Groffen, S. J., Hu, Z., Bosch, D. and Vlak, J. M. (1995). Characterization of baculovirus insecticides expressing tailored *Bacillus thuringiensis* CryIA(B) crystal proteins. *J. Invertebr. Pathol.* 66(3), 249–257.

Martínez-Barrera, O. Y., Toledo, J., Liedo, P., Gómez, J., Valle-Mora, J., Cancino, J. and Montoya, P. (2019). Does *Beauveria bassiana* (Hypocreales: Cordycipitaceae) affect the survival and fecundity of the parasitoid *Coptera haywardi* (Hymenoptera: Diapriidae)? *Environ. Entomol.* 48(1), 156–162.

Mau, R., Jang, E. B. and Vargas, R. (2007). The Hawaii area-wide fruit fly pest management programme: Influence of partnerships and a good education programme. In: Vreysen, M. J. B., Robinson, A. S. and Hendrich, J. (Eds), *Area-Wide Control of Pest Insects: From Research to Field Implementation*, Springer, The Netherlands, pp. 671–683.

Mayerhofer, J., Eckard, S., Hartmann, M., Grabenweger, G., Widmer, F., Leuchtmann, A. and Enkerli, J. (2017). Assessing effects of the entomopathogenic fungus *Metarhizium brunneum* on soil microbial communities in *Agriotes* spp. biological pest control. *FEMS Microbiol. Ecol.* 93(10), 117.

McDougall, P. (2016). The cost of new agrochemical product discovery, development and registration consultancy study. Available at: https://croplife.org/wp-content/uploads/2016/04/Cost-of-CP-report-FINAL.pdf (accessed August 2020).

McKinnon, A. C., Glare, T. R., Ridgway, H. J., Mendoza Mendoza, A., Holyoake, A., Godsoe, W. K. and Bufford, J. L. (2018). Detection of the entomopathogenic fungus *Beauveria bassiana* in the rhizosphere of wound-stressed *Zea mays* plants. Front. Microbiol. 9, 1161. doi: 10.3389/fmicb.2018.01161.

McKinnon, A. C., Saari, S., Moran-Diez, M. E., Meyling, N. V., Raad, M. and Glare, T. R. (2017). *Beauveria bassiana* as an endophyte: A critical review on associated methodology and biocontrol potential. *BioControl*. 62(1), 1–17.

Meyling, N. V. and Eilenberg, J. (2007). Ecology of the entomopathogenic fungi *Beauveria bassiana* and *Metarhizium anisopliae* in temperate agroecosystems: Potential for conservation biological control. *Biol. Control*. 43(2), 145–155.

Meyling, N. V., Lübeck, M., Buckley, E. P., Eilenberg, J. and Rehner, S. A. (2009). Community composition, host range and genetic structure of the fungal entomopathogen Beauveria in adjoining agricultural and seminatural habitats. *Mol. Ecol.* 18(6), 1282–1293.

Moore, S. and Kirkman, W. (2010). Helicovir™: A virus for the biological control of bollworm. South Afr. Fruit J. 9, 63–67.

© Burleigh Dodds Science Publishing Limited, 2022. All rights reserved.

Moscardi, F. (2007). A nucleopolyhedrovirus for control of the velvetbean caterpillar in Brazilian soybeans. In: Goettel, M. S. and Lazarovits, G. (Eds), *Biological Control: A Global Perspective*, CABI Publishing, Wallingford, UK, pp. 344-361.

Mulder, P. G., Harris, M. K. and Grantham, R. A. (2012). Biology and management of the Pecan Weevil (Coleoptera: Curculionidae). *J. Integr. Pest Manage.* 3(1), A1-A9.

Nambiar, P. T. C., Ma, S. W. and Iyer, V. N. (1990). Limiting an insect infestation of nitrogen-fixing root nodules of the pigeon pea (*Cajanus cajan*) by engineering the expression of an entomocidal gene in its root nodules. *Appl. Environ. Microbiol.* 56(9), 2866-2869.

Narciso, J. O., Ormskirk, M. M., Jones, S. A., Rolston, P., Moran-Diez, M. E., Hurst, M. R. H., Brookes, J. J. and Glare, T. R. (2019). Using multiple insecticidal microbial agents against diamondback moth larvae: Does it increase toxicity? *N. Z. J. Agric. Res.* 64, 178-193.

Nathan, S. S. and Kalaivani, K. (2006). Combined effects of azadirachtin and nucleopolyhedrovirus (SpltNPV) on *Spodoptera litura* Fabricius (Lepidoptera: Noctuidae) larvae. *Biol. Control.* 39(1), 96-104.

Nguyen, Q., Tran, T., Chan, L. C., Nielsen, L. K. and Reid, S. (2016). In vitro production of baculoviruses: Identifying host and virus genes associated with high productivity. *Appl. Microbiol. Biotechnol.* 100(21), 9239-9253.

O'Callaghan, M., Glare, T. R. and Lacey, L. A. (2012). Bioassay of bacterial entomopathogens against insect larvae. In: Lacey, L. A. (Ed.), *Manual of Techniques in Invertebrate Pathology* (2nd edn.), Academic Press, New York, pp. 101-127.

Okay, S., Tefon, B. E., Okan, M. and Ocengiz, G. (2008). Expression of chitinase A (chiA) gene from a local isolate of *Serratia marcescens* in Coleoptera-specific *Bacillus thuringiensis*. *J. Appl. Microbiol.* 104(1), 161-170.

Olson, S. (2015). An analysis of the biopesticide market now and where it is going. *Outlooks Pest Manage.* 26(5), 203-206.

Onsongo, S. K., Gichimu, B. M., Akutse, K. S., Dubois, T. and Mohamed, S. A. (2019). Performance of three isolates of *Metarhizium anisopliae* and their virulence against *Zeugodacus cucurbitae* under different temperature regimes, with global extrapolation of their efficiency. *Insects.* 10(9), 270.

Ormskirk, M. M. (2017). *Brevibacillus laterosporus as a Potential Bio-Control Agent of the Diamondback Moth and Other Insects.* PhD Thesis. Lincoln University, New Zealand.

Ormskirk, M. M., Narciso, J., Hampton, J. G. and Glare, T. R. (2019). Endophytic ability of the insecticidal bacterium *Brevibacillus laterosporus* in Brassica. *PLoS ONE* 14(5), e0216341.

Ortega, H. E., Torres-Mendoza, D. and Cubilla-Rios, L. (2020). Patents on endophytic fungi for agriculture and bio- and phytoremediation applications. *Microorganisms.* 8(8), 1237.

Ouedraogo, R. M., Goettel, M. S. and Brodeur, J. (2004). Behavioral thermoregulation in the migratory locust: A therapy to overcome fungal infection. *Oecologia.* 138(2), 312-319.

Ouna, E. A. (2010). *Entomopathogenicity of Hyphomycte Fungi to Fruit Fly Bactrocera invadens (Diptera: Tephiritidae) and Their Potential for Biological Control on Mango.* MSc Thesis. Kenyatta University, Nairobi, Kenya, p. 134.

Park, H. W., Bideshi, D. K. and Federici, B. A. (2003). Recombinant strain of *Bacillus thuringiensis* producing Cyt1A, Cry11B, and the *Bacillus sphaericus* binary toxin. *Appl. Environ. Microbiol.* 69(2), 1331-1334.

© Burleigh Dodds Science Publishing Limited, 2022. All rights reserved.

Petrika, D. T., Iseli, A.,Montelone, B. A., Van Etten, J. L. and Clem, R. J. (2003). Improving baculovirus resistance to UV inactivation: Increased virulence resulting from expression of a DNA repair enzyme. *J. Invertebr. Pathol.* 82(1), 50–56.

Prior, C., Carey, M., Abraham, Y. J., Moore, D. and Bateman, R. P. (1995). Development of a bioassay method for the selection of entomopathogenic fungi virulent to the desert locust, *Schistocerca gregaria* (Forskal). *J. Appl. Entomol.* 119(1–5), 567–573.

Qadri, M., Short, S., Gast, K., Hernandez, J. and Wong, A. C. (2020). Microbiome innovation in agriculture: Development of microbial based tools for insect pest management. *Front. Sustain. Food Syst.* 4, 547751.

Quesada-Moraga, E., Ruiz-Garcia, A. and Santiago-Alvarez, C. (2006). Laboratory evaluation of entomopathogenic fungi *Beauveria bassiana* and *Metarhizium anisopliae* against puparia and adults of *Ceratitis capitata* (Diptera: Tephritidae). *J. Econ. Entomol.* 99(6), 1955–1966.

Quesada-Moraga, E., Yousef-Naef, M. and Garrido-Jurado, I. (2020). Advances in the use of entomopathogenic fungi as biopesticides in suppressing crop pest insects. In: Birch, N. and Glare, T. (Eds), *Biopesticides for Sustainable Agriculture*, Burleigh Dodds Science Publishing, Cambridge, UK, pp. 63–98.

Rasmann, S., Köllner, T. G., Degenhardt, J., Hiltpold, I., Toepfer, S., Kuhlmann, U., Gershenzon, J. and Turlings, T. C. (2005). Recruitment of entomopathogenic nematodes by insect-damaged maize roots. *Nature.* 434(7034), 732–737.

Reardon, R. C. and Podgwaite, J. D. (1994). Summary of efficacy evaluations using aerially applied Gypchek against gypsy moth in the USA. *J. Environ. Sci. Health B.* 29(4), 739–756.

Resquín-Romero, G., Garrido-Jurado, I., Delso, C., Ríos-Moreno, A. and Quesada-Moraga, E. (2016). Transient endophytic colonizations of plants improve the outcome of foliar applications of mycoinsecticides against chewing insects. *J. Invertebr. Pathol.* 136, 23–31.

Ribeiro, B. M. and Crook, N. E. (1993). Expression of full-length and truncated forms of crystal protein genes from *Bacillus thuringiensis* subsp. *kurstaki* in a baculovirus and pathogenicity of the recombinant viruses. *J. Invertebr. Pathol.* 62(2), 121–130.

Roberts, D. W. and St. Leger, R. J. (2004). *Metarhizium* spp., cosmopolitan insect-pathogenic fungi: Mycological aspects. *Adv. Appl. Microbiol.* 54, 1–70.

Rosell, G., Quero, C., Coll, J. and Guerrero, A. (2008). Biorational Insecticides in pest management. *J. Pestic. Sci.* 33(2), 103–121.

Rudrappa, T., Czymmek, K. J., Pare, P. W. and Bais, H. P. (2008). Root-secreted malic acid recruits beneficial soil bacteria. *Plant Physiol.* 148(3), 1547–1556.

Ruiu, L. (2013). *Brevibacillus laterosporus*, a pathogen of invertebrates and a broad-spectrum antimicrobial species. *Insects.* 4(3), 476–492.

Ruiu, L., Floris, I., Satta, A. and Ellar, D. J. (2007). Toxicity of a *Brevibacillus laterosporus* strain lacking parasporal crystals against *Musca domestica* and *Aedes aegypti*. *Biol. Control.* 43(1), 136–143.

San Miguel, K. and Scott, J. G. (2016). The next generation of insecticides: dsRNA is stable as a foliar-applied insecticide. Pest Manag. Sci. 72, 801–809. doi: 10.1002/ps.4056.

Scheepmaker, J. W. A. and Butt, T. M. (2010). Natural and released inoculum levels of entomopathogenic fungal biocontrol agents in soil in relation to risk assessment and in accordance with EU regulations. *Biocont. Sci. Technol.* 20(5), 503–552.

Scheepmaker, J. W. A., Hogervorst, P. A. M. and Glandor, D. C. M. (2016). *Future Introductions of Genetically Modified Microbial Biocontrol Agents in the EU Are*

© Burleigh Dodds Science Publishing Limited, 2022. All rights reserved.

Current EU Legislation and Risk Assessment Fit for Purpose? RIVM Letter Report, Bilthoven, The Netherlands, pp. 2016-0057.

Senthilkumar, N., Murugan, K. and Zhang, W. (2008). Additive interaction of *Helicoverpa armigera* nucleopolyhedrovirus and azadirachtin. *BioControl.* 53(6), 869–880.

Shapiro-Ilan, D. I., Han, R. and Dolinksi, C. (2012). Entomopathogenic nematode production and application technology. *J. Nematol.* 44(2), 206–217.

Singhania, R. R., Patel, A. K., Thomas, L. and Pandey, A. (2016). Solid-state fermentation. In: Wittman, C. and Liao, J. C. (Eds), *Solid-State Fermentation. Industrial Biotechnology*, Wiley-VCH Verlag GmbH and Co. KGaA, Weinheim, pp. 187–204.

Soliman, N. A., Al-amin, S. M., Mesbah, A. E., Ibrahim, A. M. A. and Mahmoud, A. M. A. (2020). Pathogenicity of three entomopathogenic fungi against the Mediterranean fruit fly, *Ceratitis capitata* (Wiedemann) (Diptera: Tephritidae). *Egypt. J. Biol. Pest. Control.* 30(1), 49.

Soliman, N. A., Ibrahim, A. A., Shams El-Deen, M. M., Shams Ramadan, N. F. and Farag, S. R. (2014). Entomopathogenic nematodes and fungi as biocontrol agents for the peach fruit fly, *Bactrocera zonata* (Saunders) and the Mediterranean fruit fly, *Ceratitis capitata* (Wiedemann) soil-borne stages. *Egypt. J. Biol. Pest.* Contol. 24, 497–502.

Song, Z. (2018). Fungal microsclerotia development: Essential prerequisites, influencing factors, and molecular mechanism. *Appl. Microbiol. Biotechnol.* 102(23), 9873–9880.

Song, Z. Y., Yin, Y. P., Jiang, S. S., Liu, J. J. and Wang, Z. K. (2014). Optimization of culture medium for microsclerotia production by *Nomuraea rileyi* and analysis of their viability for use as a mycoinsecticide. *BioControl.* 59(5), 597–605.

Sorokan, A., Benkovskaya, G., Burkhanova, G., Blagova, D. and Maksimov, I. (2020). Endophytic strain *Bacillus subtilis* 26DCryChS producing Cry1Ia toxin from *Bacillus thuringiensis* promotes multifaceted potato defense against *Phytophthora infestans* (Mont.) de Bary and pest *Leptinotarsa decemlineata* Say. *Plants.* 9(9), 1115.

Sparks, T. C., Thompson, G. D., Larson, L. L., Kirst, H. A., Jantz, O. K., Worden, T. V., Hertlein, M. B. and Busacca, J. D. (1995). Biological characteristics of the spinosyns: A new naturally derived insect control agent. *Proc. Beltwide Cotton Conf.* 2, 903–907.

Spence, E. L., Chandler, D., Edgington, S., Berry, S. D., Martin, G., O'Sullivan, C., Svendsen, C. and Hesketh, H. (2020). A standardised bioassay method using a bench-top spray tower to evaluate entomopathogenic fungi for control of the greenhouse whitefly, *Trialeurodes vaporariorum*. *Pest. Manage. Sci.* 76(7), 2513–2524.

Srinivasan, R., Sevgan, S., Ekesi, S. and Tamò̀c, M. (2019). Biopesticide-based sustainable pest management for safer production of vegetable legumes and brassicas in Asia and Africa. *Pest. Manage. Sci.* 75(9), 2446–2454.

St. Leger, R., Joshi, L., Bidochka, M. J. and Roberts, D. W. (1996). Construction of an improved mycoinsecticide overexpressing a toxic protease. *Proc. Natl. Acad. Sci. USA.* 93(13), 6349–6354.

Thomas, M. B.and Jenkins, N. E. (1997). Effects of temperature on growth of *Metarhizium flavoviride* and virulence to the variegated grasshopper, *Zonocerus variegatus*. *Mycol. Res.* 101(12), 1469–1474.

Thomas, M. B., Wood, S. N. and Lomer, C. J. (1995). Biological control of locusts and grasshoppers using a fungal pathogen - the importance of secondary cycling. *Proc. R. Soc. Lond. B.* 259(1356), 265–270.

Toledo, J., Flores, S., Campos, S., Villaseñor, A., Enkerlin, W., Liedo, P., Valle, Á and Montoya, P. (2017). Pathogenicity of three formulations of *Beauveria bassiana* and efficacy of

© Burleigh Dodds Science Publishing Limited, 2022. All rights reserved.

autoinoculation devices and sterile fruit fly males for dissemination of conidia for the control of *Ceratitis capitata*. *Entomol. Exp. Appl.* 164(3), 340–349.

Tomalski, M. D. and Miller, L. K. (1992). Expression of a paralytic neurotoxin gene to improve insect baculoviruses as biopesticides'. *Nat. Biotechnol.* 10(5), 545–549.

Tseng, M. N., Chung, P. C. and Tzean, S. S. (2011). Enhancing the stress tolerance and virulence of an entomopathogen by metabolic engineering of dihydroxynaphthalene melanin biosynthesis genes. *Appl. Environ. Microbiol.* 77(13), 4508–4519.

USDA Forest Service (2020). Pesticide management & coordination. Available at: https://www.fs.fed.us/foresthealth/protecting-forest/integrated-pest-management/pesticide-management/pesticide-virus-products.shtml (accessed 1 February 2021).

van Zijll de Jong, E., Roush, T. L., Glare, T. R. and Hampton, J. G. (2016). Discovery of two *Brevibacillus laterosporus* isolates from brassica with insecticidal properties against diamondback moth. *Biocont. Sci. Technol.* 26, 426–431.

Vargas, R. I., Piñero, J. C. and Leblanc, L. (2015). An overview of pest species of *Bactrocera* fruit flies (Diptera: Tephritidae) and the integration of biopesticides with other biological approaches for their management with a focus on the Pacific Region. *Insects.* 6(2), 297–318.

Vega, F. E. (2018). The use of fungal entomopathogens as endophytes in biological control: A review. *Mycologia.* 110(1), 4–30.

Villamizar, L. F., Nelson, T. L., Jones, S. A., Jackson, T. A., Hurst, M. R. H. and Marshall, S. D. G. (2018). Formation of microsclerotia in three species of *Beauveria* and storage stability of a prototype granular formulation. *Biocont. Sci. Technol.* 28(12), 1097–1113.

Vogel, E., Santos, D., Mingels, L., Verdonckt, T. W.. and Broeck, J. V. (2019). RNA interference in insects: Protecting beneficials and controlling pests. *Front. Physiol.* 9, 1912.

Waage, J. (1997). Global developments in biological control and the implications for Europe. *EPPO Bull.* 27(1), 5–13.

Wang, H. H., Wang, J.-L., Li, Y.-P., Liu, X., Wen, J.-Z. and Lei, Z.-R. (2011a). Liquid culturing of microsclerotia of *Beauveria bassiana*, an entomopathogenic fungus to control western flower thrip, *Frankliniella occidentalis*. *J. Appl. Ent.* 48, 588–595.

Wang, S., Fang, W., Wang, C. and St. Leger, R. J. (2011b). Insertion of an esterase gene into a specific locust pathogen (*Metarhizium acridum*) enables it to infect caterpillars. *PLoS Pathog.* 7(6), e1002097.

Whipps, J. M. (1997). Developments in the biological control of soil-borne plant pathogens. *Adv. Bot. Res.* 26, 1–134.

Wraight, S. P., Jackson, M. A. and de Lock, S. L. (2001). Production, stabilization and formulation of fungal biocontrol agents. In: Butt, T. M., Jackson, C. W. and Magan, N. (Eds) Fungi as Biocontrol Agents, CABI Publishing, pp. 253–287.

Yan, S., Ren, B., Zeng, B. and Shen, J. (2020). Improving RNAi efficiency for pest control in crop species. *BioTechniques.* 68(5), 283–290.

Yousef, M., Aranda-Valera, E. and Quesada-Moraga, E. (2018). Lure-and-infect and lure-and-kill devices based on *Metarhizium brunneum* for spotted wing *Drosophila* control. *J. Pest. Sci.* 91(1), 227–235.

Yousef, M., Garrido-Jurado, I., Ruíz-Torres, M. and Quesada-Moraga, E. (2017). Reduction of adult olive fruit fly populations by targeting preimaginals in the soil with the entomopathogenic fungus *Metarhizium brunneum*. *J. Pest. Sci.* 90(1), 345–354.

Zimmermann, G., Papierok, B. and Glare, T. (1995). Elias Metschnikoff, Elie Metchnikoff or Ilya Ilich Mechnikov (1845–1916): A pioneer in insect pathology, the first describer

© Burleigh Dodds Science Publishing Limited, 2022. All rights reserved.

of the entomopathogenic fungus *Metarhizium anisopliae* and how to translate a Russian name. *Biocon. Sci. Technol.* 5(4), 527–530.

Zitlalpopoca-Hernandez, G., Najera-Rincon, M. B., del-Val, E., Alarcon, A., Jackson, T. and Larsen, J. (2017). Multitrophic interactions between maize mycorrhizas, the root-feeding insect *Phyllophaga vetula* and the entomopathogenic fungus *Beauveria bassiana*. *Appl. Soil Ecol.* 115, 38–43.

© Burleigh Dodds Science Publishing Limited, 2022. All rights reserved.

Chapter 2

Advances in bioprotectants for plant disease control in horticulture

Philippe C. Nicot, Thomas Pressecq and Marc Bardin, INRAE, Pathologie Végétale, France

1 Introduction

Horticultural crops (as defined by the International Society for Horticultural Science; https://www.ishs.org/defining-horticulture; accessed in January 2021) encompass a particularly high number of plant species which are each challenged by a large gamut of plant pathogens. Disease control is necessary to ensure adequate production and product quality. For many of the crops (ornamentals, but also fruits and vegetables), the consumer expects a high level of visual quality (possibly 'zero-defect'), particularly when the produce does not undergo substantial processing before its commercialization (Porat et al., 2018; Leus, 2018). Farmers often rely on chemical control for disease management in these crops, but bioprotectants are increasingly available and may contribute to more sustainable pest management strategies.

Among the four types of commercial bioprotectants described by the International Biocontrol Manufacturers Association (IBMA, 2019), two may be available to farmers for use against diseases of horticultural crops: microbial bioprotectants and natural substances. Microbial bioprotectants are preparations based on microorganism active substances. According to a definition jointly proposed by the Food and Agricultural Organization of the United Nations and the World Health Organization, a microorganism active

http://dx.doi.org/10.19103/AS.2021.0095.02
© Burleigh Dodds Science Publishing Limited, 2022. All rights reserved.

substance may contain viable and/or non-viable microorganisms. It can contain relevant metabolites/toxins produced during cell proliferation (growth), and material from the growth medium, provided none of these components have been intentionally altered (FAO and WHO, 2017). *Natural substances*, as defined by IBMA (2019), 'consist of one or more components that originate from nature, including but not limited to: plants, algae/micro algae, animals, minerals, bacteria, fungi, protozoans, viruses, viroids and mycoplasmas' and 'they can either be sourced from nature or are nature-identical if synthetized'. These two types of bioprotectants are sometimes referred to as 'biopesticides' (Marrone, 2019). In most countries, their commercialization is regulated by a formal registration process (FAO and WHO, 2017), in the same manner as synthetic chemical pesticides (see for example the 'List of databases on registered plant protection products in the EPPO region' provided by the European and Mediterranean Plant Protection Organization; https://www.eppo.int/ accessed in January 2021).

Considering the increasing concerns about negative side effects of chemical pesticides in many countries, bioprotectants present a favorable profile in terms of public health and environmental issues (van Lenteren et al., 2018). Their use may thus present a number of advantages for the horticulturist. In fact, it is estimated that 80% of the use of biopesticides at a global scale (both against pests and diseases) is on fruits and vegetables (Marrone, 2019). Increasingly, the commercialization of horticultural crops benefits from quality labels or eco-labels that allow the consumer to identify those produced without any, or with limited, synthetic chemical pesticides (Chen et al., 2018; Nguyen and Le, 2020; Thøgersen et al., 2019). In the production phase on the farm, regulations in numerous countries allow crop treatments with bioprotectants at periods of the growing season when protection is needed against diseases, but chemical control is restricted to limit the presence of toxic residues on the plants. This could be the case, for instance, close to harvest or postharvest and close to the selling time and consumption of the produce.

However, the protection of horticultural crops against diseases with the help of natural substances or microbial bioprotectants also presents a number of challenges, which will be addressed in the present chapter. One key challenge is a large gap between the needs of horticulturists and the limited number of bioprotectants available to them, even for major diseases of economically important crops. The issue is thus how to increase the number of registered products. Another challenge, for products that are available to farmers, is the level and consistency of the protection that can be achieved. Bioprotectants are often considered to be less efficacious and less reliable than synthetic chemical pesticides. The issues are thus to enhance the field efficacy of bioprotectants and to help farmers optimize their deployment to improve reliability.

© Burleigh Dodds Science Publishing Limited, 2022. All rights reserved.

2 Availability of bioprotectants against diseases of horticultural crops

One feature of horticultural crops is the very large number of plant species they include, most of which are considered as 'minor' crops on the basis of their cultivation area, production and consumption (SANCO, 2017; OECD, 2009). Many different plant pathogens may cause diseases in those crops, and most of these pathogens have narrow host ranges. This results in a vast range of possible host–pathogen combinations, each corresponding to a potential need for a plant protection solution.

The number of host–pathogen combinations (or 'uses') for which at least one registered bioprotectant is available to horticulturists may vary greatly from one country to another, depending on national regulations. However, it is likely to represent a small fraction of all possible needs by farmers. For example, among bioprotectants registered in France, microbial biocontrol agents covered 163 uses in 2020 (Table 1), with 36 commercial products based on 26 microbial preparations (16 with fungal strains, 8 with bacterial strains and 2 with viruses). Most of these products are registered for application on above-ground plant parts in the field or after harvest (Table 1A), but several may be used for treatment of seeds and transplants (Table 1B) and a few are registered for the treatment of soil or growth substrates (Table 1C).

Without considering sulfur- and copper-based products, bioprotectants based on natural substances cover 61 horticultural uses in France (Table 2), 34 of which are in common with microbial biocontrol agents. These natural substances are mostly plant or microbial extracts, but the majority of commercial products are based on phosphonates (20 products) and potassium carbonate (3 products). In France, several natural substances (and one microbial bioprotectant) benefited from a general registration for use as 'stimulation of plant defenses', without the mention of any specific disease or plant pathogen. Sulfur-based active substances (55 commercial products) cover 41 uses, mostly against powdery mildew of different crops. Copper-based active substances (10 products) cover 54 uses, mostly against downy mildews or late blight and bacterial diseases (Table 2). Overall, bioprotectants registered in France in 2020 (if one includes sulfur- and copper-based products) cover 244 distinct uses on horticultural crops, of which 31 are specific to copper-based products and 13 to sulfur-based products.

Despite these three-digit figures on uses covered by bioprotectants, which tend to increase year after year, a large part of the potential needs of horticulturists is not well covered at a global scale. In fact, many of the needs are not covered by any plant protection product at all, complicating the management of diseases by farmers. The situation has been made more severe by the evolution of the safety standards set by regulatory authorities in the last

© Burleigh Dodds Science Publishing Limited, 2022. All rights reserved.

Table 1 Examples of microbial strains used as active substances in bioprotectants registered in France for disease control on horticultural crops (from the E-phy database https://ephy.anses.fr/ consulted in October 2020)

Microbial strains	Registered use: target disease/pathogen and crops
A: Products which can be applied as foliar sprays or post-harvest treatments	
Ampelomyces quisqualis (strain AQ10)	**Powdery mildew** (on cucumber, melon, pepper, strawberry, tomato, rose)
Aureobasidium pullulans (strains DSM 14940 & DSM 14941)	**Gray mold** (grapes, strawberry, tomato); **fireblight** (apple); **post-harvest** (apple, citrus)
Bacillus amyloliquefaciens (strain FZB24)	***Botrytis* and *Sclerotinia*** (lettuce); **brown spot** (potato, tomato); **downy mildew or late blight** (cucumber, grapes, lettuce, potato, tomato); **gray mold** (grapes); **powdery mildew** (cucumber, grapes, tomato); **unspecified bacteria** (tomato)
Bacillus amyloliquefaciens (strain MBI600)	***Botrytis* and *Sclerotinia*** (lettuce, pepper, strawberry, tobacco, tomato); **gray mold** (grapes, raspberry); **unspecified fungi** (chicory, mushroom)
Bacillus amyloliquefaciens ssp. plantarum (strain D747)	***Botrytis* and *Sclerotinia*** (cucumber, lettuce, melon); **fireblight** (apple); **gray mold** (blackcurrant, grapes, pepper, raspberry, strawberry, tomato); ***Monilinia* brown rot** (cherry, peach, plum); **post-harvest rot** (apple); **powdery mildew** (blackcurrant, pepper, raspberry, strawberry, tomato); **Stemphylium** (apple); **unspecified bacteria** (kiwi)
Bacillus pumilus (strain QST 2808)	**Powdery mildew** (aromatic and medicinal plants, black currant, chicory, cucumber, flowers and potted plants, grapes, herbs, lettuce, melon, pepper, raspberry, strawberry, tobacco, tomato, vegetable crops)
Bacillus subtilis (strain IAB/BS03)	**Powdery mildew** (cucumber, lettuce); **scab** (apple)
Bacillus subtilis (strain QST 713)	Bacterial diseases: **fire blight** (apple, pear) and **various bacteria** (almond, aromatic and medicinal plants, artichoke, cardoon, cherry, eggplant, jujube, kiwi, nectarine, nuts, olive, ornamental plants, peach, pepper, plum, tomato) Fungal diseases: **anthracnose** (avocado); ***Botrytis* and *Sclerotinia*** (beans, chicory, cowpea, courgette, cucumber, eggplant, fava beans, lamb's lettuce, lentils, lettuce, melon, peas, pepper, rocket, strawberry, tobacco, tomato, watermelon); **gray mold** (aromatic and medicinal plants, blackberry, blackcurrant, grapes, lingonberry, ornamental plants, raspberry, tobacco, tomato); ***Monilinia* brown rot** (almond, cherry, jujube, nectarine, peach, plum; stone fruits); **Mycosphaerella leaf streak** (banana); **peacock eye** and other olive fruit diseases (olive); **powdery mildew** (aromatic and medicinal plants, blackberry, blackcurrant, cucumber, flowers and potted plants, lingonberry, mango, melon, ornamental plants, raspberry, rose, small fruits, trees and shrubs); **scab** (apple); ***Sclerotinia* collar rot** (lettuce); **Stemphylium** (pear); **unspecified fungi** (aromatic and medicinal plants) Post-harvest rot (kiwi)

(*Continued*)

© Burleigh Dodds Science Publishing Limited, 2022. All rights reserved.

Table 1 (*Continued*)

Microbial strains	Registered use: target disease/pathogen and crops
Candida oleophila (strain O)	**Post-harvest rot** (apple, banana, citrus)
Clonostachys rosea (*Gliocladium catenulatum*) (strain J1446)	***Botrytis* and *Sclerotinia*** (cucumber, flowers and green plants, pepper, raspberry, strawberry, tomato) and **other fungal diseases on leaves** (cucumber, flowers and potted plants, pepper, tomato)
Metschnikowia fructicola (strain NRRL Y-27328)	***Botrytis* and *Sclerotinia*** (blackcurrant, grapes, raspberry, strawberry); ***Monilinia* brown rot** (cherry, peach, plum)
Phlebiopsis gigantea	**Armilaria** on tree stumps
Saccharomyces cerevisiae (strain LAS02)	***Botrytis* and *Sclerotinia*** (blackcurrant, grapes, pepper, raspberry, strawberry, tomato); ***Monilinia* brown rot** (peach, plum)
Trichoderma atroviride (strain SC1)	***Botrytis* and *Sclerotinia*** (tomato, grapes); **wood-decay fungi** (grapes)
Trichoderma atroviride (strain I-1237)	**Oomycetes** (carrot); **unspecified fungi** (lettuce); **wood-decay fungi** (grapes)
Trichoderma gamsii (strain ICC080) + *Trichoderma asperellum* (strains ICC012 T25 & TV1)	**Wood-decay fungi** (grapes)
B: Products which can be applied to seeds and nursery transplants	
Bacillus amyloliquefaciens (strain MBI600)	**Unspecified fungi** (witloof chicory)
Pseudomonas sp. (strain DSMZ 13134)	**Unspecified fungi** (potato)
Trichoderma asperellum (strain T34)	**Wilt diseases** (ornamental plants)
Trichoderma atroviride (strain I-1237)	**Unspecified fungi** (potato)
Trichoderma atroviride (strain SC1)	**Wood-decay fungi** (grapes)
Trichoderma harzianum Rifai (strains T-22 & ITEM-908)	**Unspecified fungi** (aromatic and medicinal plants)
Mild pepino mosaic virus (isolates VC1 & VX1)	**Pepino mosaic virus** (tomato)
Pepino Mosaic Virus (strain CH2, isolate 1906)	**Pepino mosaic virus** (tomato)
C: Products which can be applied as soil or substrate treatments	
Bacillus amyloliquefaciens (strain MBI600)	**Unspecified fungi** (mushroom)
Bacillus amyloliquefaciens *ssp. plantarum* (strain D747)	**Substrate disinfestation** (mushroom)

(*Continued*)

© Burleigh Dodds Science Publishing Limited, 2022. All rights reserved.

Table 1 (*Continued*)

Microbial strains	Registered use: target disease/pathogen and crops
Bacillus subtilis (strain QST 713)	**Soil oomycetes** (carrot); **various soil fungi** (cucumber, lettuce, melon, pepper, potato, tomato); **competing fungi and substrate disinfestation** (mushroom)
Clonostachys rosea (*Gliocladium catenulatum*) (strain J1446)	**General soil treatment** against soil fungi and oomycetes
Paraphaeosphaeria minitans (*Coniothyrium minitans*) (strain CON/M/91-08)	**General soil treatment** against *Sclerotinia*
Pseudomonas sp. (strain DSMZ 13134)	**Various soil fungi** (aromatic and medicinal plants, beet, cabbage, chicory, cucumber, lettuce, melon, pepper, poppy, potato, spinach, strawberry, tomato, turnip, watercress)
Streptomyces griseoviridis (strain K61)	**General soil treatment** against soil fungi and oomycetes
Trichoderma asperellum (strain T34)	**Wilt-causing fungi** (ornamentals); **soil oomycetes** (pepper, tomato)
Trichoderma asperellum(strains ICC012 T25 & TV1)	**Soil oomycetes** (vegetable crops); **general soil treatment** against soil fungi and oomycetes
Trichoderma asperellum (strains ICC012 T25 & TV1) + *Trichoderma atroviride* (formerly *T. harzianum*) (strainT11)	**Soil fungi and oomycetes** (trees and shrubs, cucumber, flowers and potted plants, melon, ornamental crops, pepper, vegetable crops, tomato); **soil oomycetes** (strawberry); **soil fungi** (lettuce)
Trichoderma atroviride (strain I-1237)	**Soil oomycetes** (carrot); **soil fungi** (ornamental crops, lettuce, potato)
Trichoderma gamsii (strain ICC080) + *Trichoderma asperellum* (strains ICC012 T25 & TV1)	**Armilaria** (ornamental crops); **soil oomycetes** (ornamental and vegetable crops); **soil fungi** (strawberry); **soil disinfestation** (fruit trees)
Trichoderma harzianum Rifai (strains T-22 & ITEM-908)	**General soil treatment** against soil fungi and oomycetes; **soil fungi and oomycetes** (carrot, cabbage, cauliflower, broccoli, cucumber, ornamental crops, herbs, strawberry, grasses, beans, lettuce, turnip, onion, melon, leek, pepper, tomato); **soil oomycetes** (celery, fruit and vegetable crops); **soil fungi** (aromatic and medicinal plants)

decades. The reassessment of toxicological issues with previously registered plant protection products has resulted in a drastic reduction in the number of synthetic chemical pesticides that are authorized for commercialization in the European Union and many other countries (Donley, 2019). The problems posed by requirements for which no plant protection product is available have long been a concern for farmers, including horticulturists, and are dealt with by expert

© Burleigh Dodds Science Publishing Limited, 2022. All rights reserved.

groups at national and international levels. One example is the European Union Minor Uses Co-ordination Facility, which maintains a database compiling all unfulfilled needs reported by EU member states. Horticultural crops represent a high proportion of recorded needs, amounting to 2862 of the 3177 minor uses compiled in this database (http://www.eumuda.eu/index; accessed in October 2020). The great majority of these needs are for vegetable crops (covering over 100 plant species) and for fruits and nuts (over 50 species) (Fig. 1). For all horticultural crops, plant diseases account for 34% of the reported needs, with animal pests and weeds each representing 52% and 14%, respectively. The range of implicated pathogens of horticultural crops is wide, encompassing over 300 species of fungi or oomycetes, 30 species or pathovars of bacteria and 24 viruses.

Thus, there is a globally strong need for more active substances to be registered as bioprotectants for a wide range of uses.

3 Advances in increasing the number of bioprotectants for disease control

Several approaches may contribute to increasing the number of active substances registered as bioprotectants for a wider range of horticultural uses.

3.1 Extending the spectrum of activity of existing products

Among bioprotectants currently available for horticulturists, most cover a limited number of uses, as illustrated for France in Tables 1 and 2. This situation may be due to the inability of the registered microbial strains or natural substances to provide plant protection beyond a narrow range of host-pathogen combinations. This is the case, for example, for the mycoparasite *Ampelomyces quisqualis*, whose biological activity is limited to the fungi responsible for powdery mildew diseases (Pintye et al., 2015).

However, it may also be due to a lack of information on the spectrum of activity of the bioprotectants, as the acquisition of such knowledge requires time-consuming and costly experimental research and field trials. Over the years, the range of registered uses for certain bioprotectants has markedly increased, possibly in relation with the growth of the original biocontrol companies that developed them or with the increasing involvement of large international plant protection businesses with strong investment capacities in research and development (Marrone, 2019; Pelaez and Mizukawa, 2017). For example, the strain QST713 of *Bacillus subtilis*, originally registered by the company Agraquest in the USA, was commercialized in France with a single use in 2006 (Table 3). In 2020, its registration in France covered over 60 uses on horticultural crops and several on arable crops.

© Burleigh Dodds Science Publishing Limited, 2022. All rights reserved.

Table 2 Examples of natural substances used in bioprotectants registered in France for disease control on horticultural crops (from the E-phy database https://ephy.anses.fr/ consulted in October 2020)

Natural substance[a]	Registered use (target disease/pathogen and crops)[b]
Cerevisiane (2 products)	**Stimulation of plant defenses** (cucumber, strawberry, lettuce, melon, tomato, grapes)
Clove oil + plant oils	**Post-harvest rot** of apple
Copper (54 uses ; 10 products)	**Anthracnose** (hazelnut); **Bacteria** (apple, artichoke, bean, cabbage, celery, chicory, grapes, kiwi, leek, lettuce, melon, nut trees, olive, onion, peach, pea, plum, strawberry, tomato); **brown spot** (strawberry); **Coryneum blight and Polystigma red leaf spot** (cherry, peach, plum); **downy mildew and late blight** (artichoke, cabbage, cucumber, grapes, leek, lettuce, melon, onion, potato, potted plants, rose, tomato, trees & shrubs); **fungal canker** (almond); **fungal dieback** (hazelnut); **leaf curl** (cherry, peach, plum); **leaf spots** (potted plants, trees & shrubs); **Mycosphaerella leafspot** (chestnut); **Nectria canker** (apple); **oomycetes** (carrot); **peacock eye** (olive); **scab** (apple); **various unspecified diseases** (citrus)
COS-OGA (2 products)	**Stimulation of plant defenses** (grapes, vegetable crops)
Essential oil from orange (2 products)	**Black spot** (rose); **brown spot** (tomato); **downy mildew or late blight** (cucumber, flowers & potted plants, grapes, lettuce, melon, rose, tobacco, tomato, turnip); **powdery mildew** (apple, blackcurrant, carrot, chicory, cucumber, flowers & potted plants, grapes, herbs, hops, lettuce, mango, melon, papaya, peach, rose, raspberry, salsify, strawberry, tobacco, tomato, trees and shrubs); **rusts** (cauliflower, flowers & potted plants, rose, trees and shrubs); **unspecified fungi** (herbal tea plants)
Fenugreek extract	**Stimulation of plant defenses** (grapes, melon)
Laminarin	**Fire blight** (apple); **powdery mildew** (strawberry); **scab** (apple); **stimulation of plant defenses** (apple, grapes, lettuce, strawberry)
Potassium carbonate (3 products)	**Gray mold** (grapes); ***Monilinia* brown rot** (peach); **powdery mildew** (apple, artichoke, blackcurrant, cucumber, flowers & potted plants, grapes, hops, melon, pepper, raspberry, rose, strawberry, tomato, trees and shrubs); scab (apple); **unspecified fungi** (aromatic & medicinal plants)
Potassium or sodium phosphonates (20 products)	**Collar rot and oomycetes** (citrus, fruit trees); **downy mildew** (grapes); **scab** (apple); **stimulation of plant defenses** (chicory, cucumber, lettuce, melon, pepper, tomato) + **post-harvest** (citrus) + **soil treatment against collar rot** of fruit trees
Sulfur (42 uses ; 55 products)	**Black spot** (rose); **Cercospora leaf spot** (beet); **downy mildew** (grapes); **leaf spots** (potted plants); **Phomopsis cane and leaf spot** (grapes); **powdery mildew** (apple, artichoke, beans, beet, blackcurrant, carrot, celery, chicory, cucumber, flowers & potted plants, garden legumes, grapes, hazelnut, hops, lettuce, mango, melon, papaya, peach, peas, pepper, rose, raspberry, salsify, spinach, strawberry, tomato, trees & shrubs); **scab** (apple); **unspecified fungi** (aromatic & medicinal plants)
Thymol	**Gray mold** (grapes)

[a] Commercialized as one single product unless otherwise specified.
[b] Products are applicable as foliar sprays unless otherwise stated.

© Burleigh Dodds Science Publishing Limited, 2022. All rights reserved.

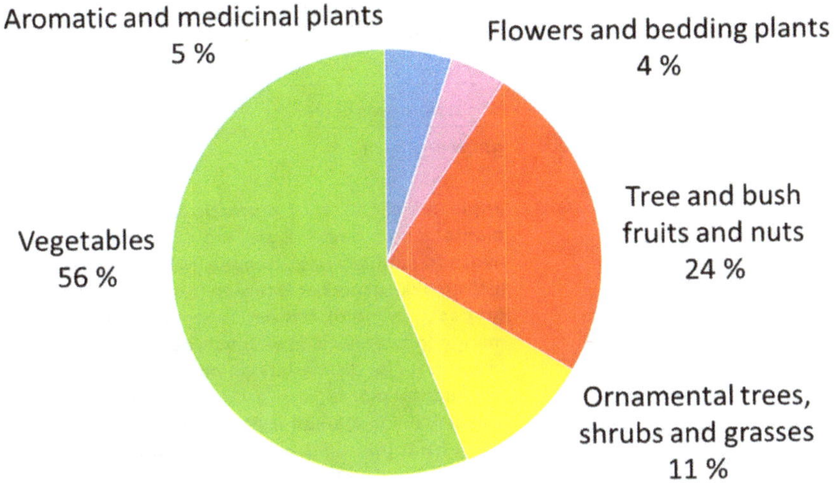

Figure 1 Distribution, according to main types of crops, of horticultural uses compiled in the 'Minor Uses Table of Needs' of the European Union Minor Uses Co-ordination Facility (accessed in October 2020).

For many existing bioprotectants, further progress may be expected on the identification of new protective effects against diseases of horticultural crops. This trend could be stimulated by information already available for the active substances of existing products at a global scale, including those with current uses registered on non-horticultural crops or in different regions of the world with various climates and agricultural practices. A recent review of microbial biological control agents registered in Australia, Brazil, Canada, the European Union, Japan, New Zealand and the USA (van Lenteren et al., 2018) shows a potential for such data mining. It presents a number of microbial strains (26 bacteria, 31 fungi and 2 viruses) and 28 species that are not listed in Table 1 of the present chapter (Fig. 2). Published scientific literature is generally available for registered strains, or for strains of similar species, and may provide useful guidance for the expansion of registered uses. It could provide evidence or assist in devising hypotheses on the protective potential of registered strains on new plant-pathogen combinations. Hypotheses could, for instance, result from careful extrapolations of known protective activity of a strain on horticultural crops to that of closely related strains of the same species. Advanced knowledge on the modes of action of a bioprotectant may also help to identify additional target pathogens or crop-pathogen combinations that could be responsive to the same protective effects.

One promising development of bioprotectants currently registered against invertebrate pests, especially when their mode of action relies on the stimulation of plant defenses, could also result from the evaluation of possible protective

© Burleigh Dodds Science Publishing Limited, 2022. All rights reserved.

Table 3 Recent evolution of uses for bioprotectants. Example of *Bacillus subtilis* QST 713 in France (sources: Index Phytosanitaire ACTA. ACTA Editions, France & the E-phy database https://ephy.anses.fr/ consulted in October 2020)

Commercialization[a]	Year	Registered uses in France[b]
Nufarm SAS (Serenade Biofungicide)	2006	**grapes** (gray mold)
BASF Agro SAS (Serenade Max)	2010	**apple** (fire blight, scab); **aromatic & medicinal plants** (various bacteria, gray mold, powdery mildew); **avocado** (anthracnose); **banana** (Mycosphaerella leaf streak); **cucumber & melon** (powdery mildew); **grapes** (gray mold); **lettuce** (*Sclerotinia* collar rot); **mango** (powdery mildew); **mushrooms** (treatment of substrate for disinfestation); **nuts** (bacteria); **ornamental plants** (various bacteria, gray mold, powdery mildew) **small fruits** (powdery mildew); **stone fruits** (*Monilinia* brown rot); **tomato** (bacteria, gray mold)
Bayer Crop Science (Serenade Max)	2018	2010 uses + **almond** (bacteria, *Monilinia* brown rot); **apple & pear** (fire blight); **aromatic and medicinal plants** (various bacterial and fungal diseases); **artichoke & cardoon** (bacteria); **blackberry, blackcurrant, lingonberry & raspberry** (gray mold; powdery mildew); **carrot & parsley** (soil treatment against oomycetes); **cherry** (bacteria, *Monilinia* brown rot); **chicory, lamb's lettuce, lettuce & rocket** (*Botrytis* and *Sclerotinia*; soil treatments against soil-borne fungi); **cucumber** (*Botrytis* and *Sclerotinia*); **eggplant & tomato** (bacteria, *Botrytis* and *Sclerotinia*, soil treatments against soil-borne fungi); **flowers and potted plants** (powdery mildew); **mushrooms** (treatment of substrate against competing fungi); **peach & nectarine** (bacteria, *Monilinia* brown rot); **pear** (Stemphylium); **pepper** (bacteria, *Botrytis* and *Sclerotinia*, soil treatment against soil-borne fungi); **plum & jujube** (bacteria, *Monilinia* brown rot); **rose** (powdery mildew); **strawberry** (*Botrytis* and *Sclerotinia*); trees & shrubs (powdery mildew); **table grapes** (gray mold); **tobacco** (*Botrytis* and *Sclerotinia*); **walnut** (bacteria)
Bayer Crop Science (Rhapsody, Serenade ASO, Serenade Max)	2020	2018 uses + **beans, cowpeas, fava beans, lentils & peas** (*Botrytis* and *Sclerotinia*), **courgette** (*Botrytis* and *Sclerotinia*); **courgette & cucumber** (soil treatment against soil-borne fungi); **kiwi** (bacteria, post-harvest rot); **melon & watermelon** (*Botrytis* and *Sclerotinia*; soil treatment against soil-borne fungi); **olive** (bacteria, peacock eye and various fruit diseases)

[a] Company (+ name of commercial product).
[b] Registered uses are for foliar application unless otherwise stated. All uses registered in 2010 were for stimulation of plant defenses, with the exception of substrate treatment for mushroom production.

© Burleigh Dodds Science Publishing Limited, 2022. All rights reserved.

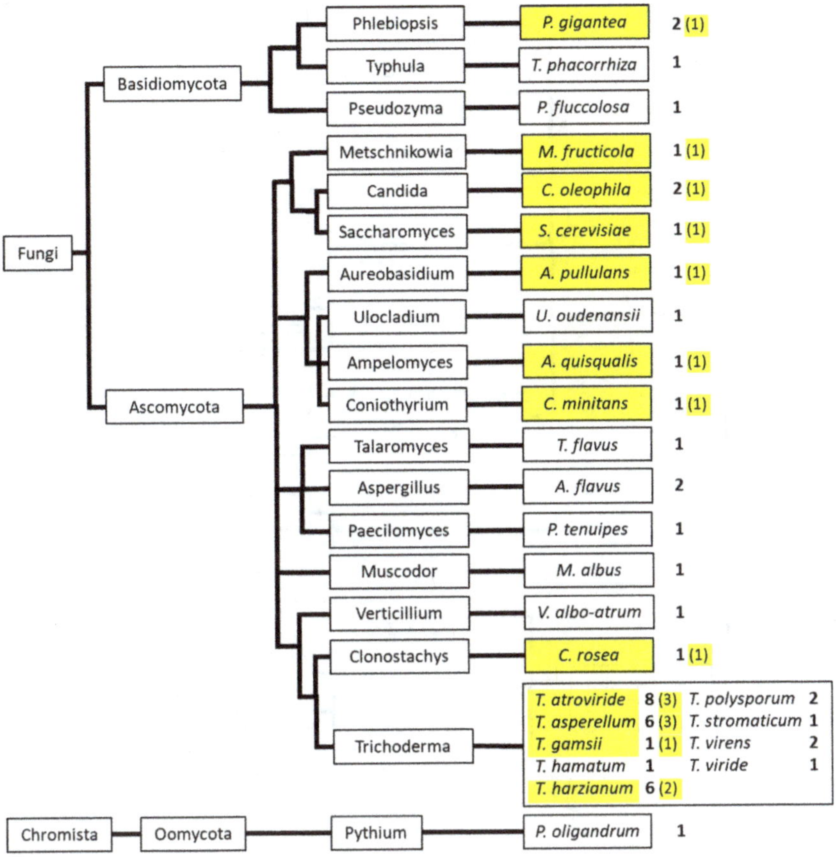

Figure 2 Diversity of genera and species of microorganisms registered as bioprotectants in Australia, Brazil, Canada, the European Union, Japan, New Zealand and the USA, based on the review of (van Lenteren et al., 2018). The dendrograms show key taxonomic links between genera. Yellow highlights indicate species registered in France on horticultural crops (listed in Table 1). For each species, bold-faced numbers show the total counts of registered strains (the subset of those listed in Table 1 is indicated between brackets).

effects against diseases. One plant extract listed in Table 2 (essential oil from orange) is also registered in various countries for the protection of numerous horticultural crops against pests such as leafhoppers, psyllids, scale insects, thrips, white flies, and mites. Several species of *Trichoderma*, a genus mostly known for its bioactivity against plant diseases, have been reported to harbor strains that can protect plants against aphids, thrips and caterpillars (Coppola et al., 2019). Vice versa, microbial species known for their entomoparasitic effects were shown to have the ability to protect horticultural crops against plant pathogens. This is the case, for example, of *Metarhizium anisopliae* against gray mold of tomato (Sarven et al., 2020) and *Verticillium lecanii* against

© Burleigh Dodds Science Publishing Limited, 2022. All rights reserved.

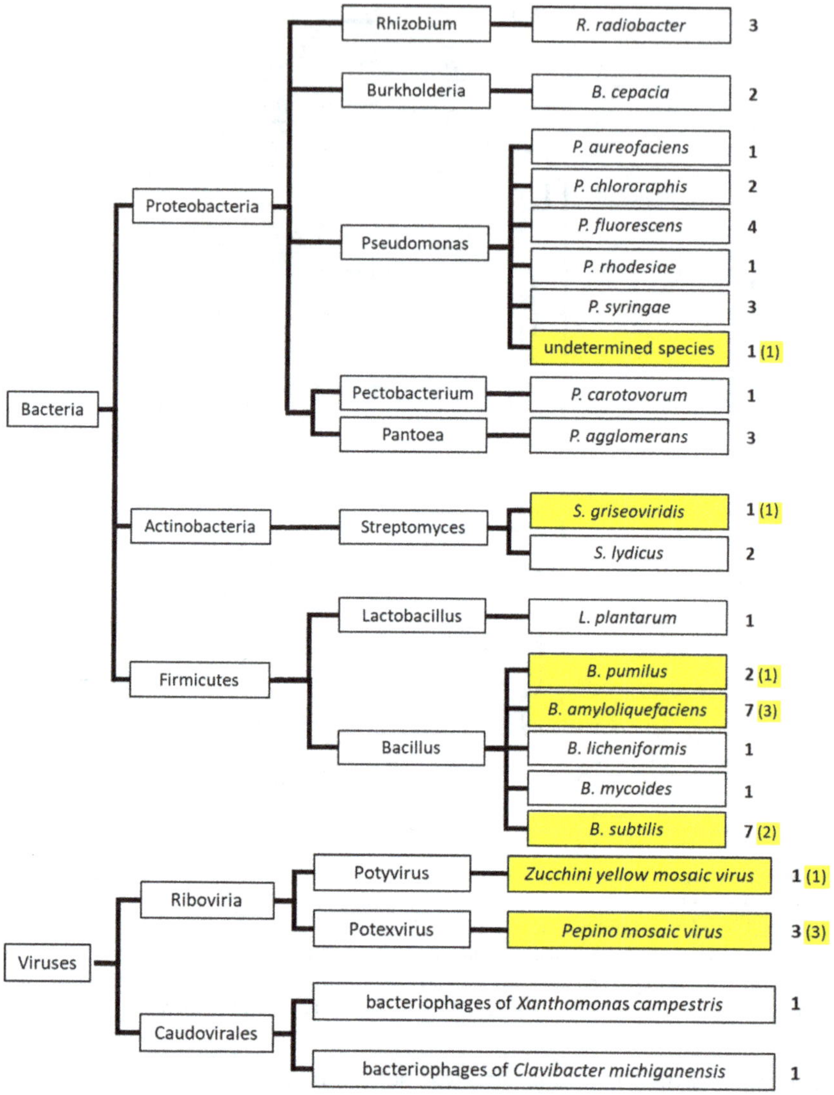

Figure 2 (*Continued*).

several powdery mildew fungi, rust fungi and *Penicillium digitatum*, the agent of post-harvest green mold on fruits (Benhamou, 2004).

3.2 Finding new active substances: where and how?

There is a high potential for escalating the discovery of new active substances with protective effects against diseases of horticultural crops. Two approaches

© Burleigh Dodds Science Publishing Limited, 2022. All rights reserved.

may stimulate advances in this domain: (i) expanding the sourcing of new candidate microbial strains or natural substances and (ii) rationalizing and upscaling screening methods for the selection of candidates.

The microorganisms currently registered as bioprotectants are limited to a small number of species and genera, suggesting that the range of sourcing for new strains could be expanded. For example, among the 47 fungal strains and 43 bacterial strains accounted for in Fig. 2, 28 belong to the genus *Trichoderma* (nearly 60% of fungal strains), 18 to *Bacillus* and 12 to *Pseudomonas* (nearly 42% and 28% of bacterial strains, respectively). In contrast, estimates of global microbial biodiversity on earth are usually in the range of millions of species (Vitorino and Bessa, 2018). A wider exploration of this vast pool of potential bioprotectants has been stimulated by the recent surge of studies on microbiomes that were fostered by advances in multiple 'omics' tools and bioinformatics (Massart et al., 2015). In the agricultural context, a special focus is placed on soil- or plant-associated microbiota (Berg et al., 2017; Ab Rahman et al., 2018; Schlatter et al., 2017; Massart et al., 2015; Compant et al., 2019). Concretely, the large amounts of generated data may present challenges, and new methods need to be devised for the interpretation of microbiome networks with the objective of identifying candidate microbes that affect plant health (Poudel et al., 2016).

Another issue with the sourcing of new bioprotectants is the choice of ecological niches where microbiomes are most likely to be interesting. Formerly, recommendations to seek potential bioprotectant strains were to focus on agricultural environments, in areas where the disease caused by a given pathogen did not occur or had declined despite the presence of a susceptible host, rather than on areas where the disease occurred (Ellis, 2017). This applies well, for example, to the study of microorganisms in suppressive soils (Schlatter et al., 2017). These early recommendations have been questioned and a wider range of possible niches is increasingly being considered, including endophytic microbiota and microbiota on diseased plants, on survival structures of the plant pathogens and in river or marine environments (Ellis, 2017; Mülner et al., 2019; Sahu et al., 2020; Hu et al., 2019; Kong, 2018; Marimuthu et al., 2020; De Silva et al., 2019). Marine samples, consisting of fish, shellfish and shrimp, as well as intertidal sand and mud collected along the seashore, proved to be a valuable source of actinomycetes for the control of blue mold of citrus (Hu et al., 2019). Exploring wider ranges of climatic conditions may also be valuable. Psychrotrophic yeasts were successfully selected from water and soil samples collected in Antarctica, for use as biocontrol agents for the post-harvest diseases of apples stored at low temperature (Arrarte et al., 2017).

Much effort is also dedicated to the exploration of new sources of natural substances, for example as extracts from a variety of terrestrial or sea plants (Raveau et al., 2020; Aguilar-Veloz et al., 2020) or from a wide variety

© Burleigh Dodds Science Publishing Limited, 2022. All rights reserved.

of microorganisms. Marine-derived bacterial lipopeptides were shown to inhibit gray mold caused by *Botrytis cinerea* on potted cucumber plants and on Chinese rose (*Rosa chinensis*) in field experiments (Gu et al., 2017). Natural products from mangrove rhizospheric microorganisms were recently reported to provide control of anthracnose, caused by *Colletotrichum fructicola* on pear fruits (Hao et al., 2019). The microbiome studies mentioned may also lead to the identification of natural substances of microbial origin (Hao et al., 2019).

Improved screening strategies and advanced phenotyping tools are also likely to accelerate the selection and industrial development of new bioprotectants. In many cases, screening of candidate microorganisms or natural substances was initially focused on criteria that pertain to their protective efficacy against one or several plant pathogens. For microorganisms, the bioassays in early studies generally relied on *in vitro* confrontation, on nutrient agar media, of microbial candidate strains with a plant pathogen to assess their biocontrol potential. This method presents many logistical advantages. Its implementation in the laboratory is simple, fast, cheap and compatible with high throughput screening. However, it focuses the selection criteria on antagonism by the candidate strains against the pathogen and the production of antimicrobial compounds, while disregarding possible protective effects that could result from interactions with the plant to be protected or with its microbiome (Berg et al., 2017). Additionally, there has been mounting evidence for a possible lack of correlation between the *in vitro* performance of a strain and its subsequent protection efficacy on the host plants, leading to the conception of more sophisticated screening procedures that implicate *in vivo* tests on plants or detached plant organs (Bouaoud et al., 2018; Raymaekers et al., 2020; Köhl et al., 2020; Besset-Manzoni et al., 2019; Köhl et al., 2019b). Increased throughput for *in vivo* tests can be expected with the involvement of automated phenotyping platforms, possibly assisted by artificial intelligence (Ozyilmaz, 2020). The evaluation of complex microbial communities may also benefit from high throughput systems such as the kChip screening platform (Kehe et al., 2019).

Aside from criteria concerning the protective efficacy of candidate microorganisms or natural substances, many other qualities need to be evaluated in a screening process leading to the industrial and commercial development of a bioprotectant. They may include a variety of biological and ecological traits such as UV and drought resistance, cost-effective mass production and compatibility with other bioprotectants or chemical pesticides. These additional selection criteria may address requirements related to industrial production and to the registration of a future product. Integrative screening strategies have been proposed to prioritize this whole broad range of criteria in relation to their cost and logistical weight (Köhl et al., 2011; Marrone, 2019; Köhl et al., 2019b).

© Burleigh Dodds Science Publishing Limited, 2022. All rights reserved.

3.3 Facilitating registration of bioprotectants

In most countries, the commercialization of a bioprotectant necessitates its registration with regulatory authorities. For this, the company needs to provide a dossier with extensive information covering the chemical, physical and biological properties of the product or active substance, as well as toxicological and environmental issues (Gotor-Vila et al., 2017; Gu et al., 2017; Marrone, 2019; Korsten, 2004). In many countries, an additional dossier must be submitted to document the field efficacy of the product (Frederiks and Wesseler, 2019). Overall, the procedure may be complex, it requires costly investments for the companies, and it generally involves a lengthy evaluation period by the regulatory authorities (Marrone, 2019; Frederiks and Wesseler, 2019). This has often been viewed as an impediment to the development of bioprotectants, particularly by small companies or when the product targets a small market (Zaki et al., 2020; Robin and Marchand, 2019; Korsten, 2004; Bailey et al., 2010; Glare et al., 2012; Chandler et al., 2011; Abbey et al., 2019; Aguilar-Veloz et al., 2020). Improvement of this situation can be expected when regulatory authorities provide incentives, such as preferential reduced administrative fees or a reduced delay for dossier evaluation, that may be granted for the registration of bioprotectants by small companies (Marrone, 2019). However, an evolution of the data requirements for the registration of bioprotectants remains to be carefully considered in many countries, as they were initially devised for the registration of chemical pesticides and may not be fully adapted for biological compounds (Köhl et al., 2019a,b; Koch et al., 2018; Chandler et al., 2011). Developing fully adapted toxicological criteria for the registration of bioprotectants to be used for post-harvest protection of horticultural crops may be particularly challenging (Aguilar-Veloz et al., 2020).

4 Enhancing the field efficacy of bioprotectants on horticultural crops

The field performance of bioprotectants is usually compared to, and expected by the farmers, to match that of chemical pesticides (Marian and Shimizu, 2019). In past decades, however, commercialized bioprotectants have shown success, but also partial success, failures and lack of economic viability on the farm, which have affected the perception of end-users about their protective efficacy (Glare et al., 2012). This assessment fits well with reviews of the scientific literature, which have shown a gap between an abundance of reports on high protective efficacy observed in the laboratory in controlled conditions, and a smaller number of reports on effective protection in the field (Nicot et al., 2012; Ojiambo and Scherm, 2006; Marian and Shimizu, 2019). Several advances, which contribute to improving this situation, will be discussed below.

© Burleigh Dodds Science Publishing Limited, 2022. All rights reserved.

4.1 Expanding knowledge on their modes of action

Knowledge on the modes of action of a bioprotectant is essential, because they affect its protective efficacy, its efficiency of use, consistency of response, and host target and non-target susceptibility (Hubbard et al., 2014; Köhl et al., 2019b; Aguilar-Veloz et al., 2020). In terms of industrial production of the bioprotectants, they may also influence the choice of manufacturing method and the final cost of production (Aguilar-Veloz et al., 2020;, Hubbard et al., 2014). Information on the modes of action may also be required in registration dossiers of bioprotectants (Köhl et al., 2019b).

The modes of action of bioprotectants can involve a direct impact on the pathogen, through antibiosis (by means of non-volatile or volatile antimicrobial compounds), competition or hyperparasitism (Fig. 3). Competition and hyperparasitism (the direct competitive interaction between two organisms in which one organism [the bioprotectant] is gaining nutrients from the other [the plant pathogen]), are specific to bioprotectants based on living microorganisms. The action of bioprotectants can also result from indirect effects through the stimulation of plant defenses or through impacts on the microbiota of the plant or the soil (Guimarães et al., 2020; Gao et al., 2017; Elsayed et al., 2020; Antoniou et al., 2017). There are excellent reviews describing all these modes of action (Köhl et al., 2019b; Singh et al., 2020; De Silva et al., 2019; Aguilar-Veloz et al., 2020; Ab Rahman et al., 2018; Rao et al., 2019) and thus they will not be presented in detail here.

One important property of bioprotectants, particularly those based on living microorganisms, is that their protective effect is rarely limited to a single mode of action. This is exemplified by species of *Trichoderma*, which have long been studied as biocontrol agents on horticultural crops (Verma et al., 2007). All of the modes of action included in Fig. 3 have been reported for these species, and a number of strains are known for their multiple modes of actions (Zaki et al., 2020; Sood et al., 2020; Verma et al., 2007). Their stimulatory effects on plant defenses implicate complex mechanisms. These can result in a localized or a systemic plant response, and depending on strains and host plants, they may mediate directly-induced defenses or defense priming (Zaki et al., 2020; Conrath et al., 2015; Mauch-Mani et al., 2017; Lorito et al., 2010).

Natural substances are often used as antimicrobial compounds, possibly also through a volatile effect (Aguilar-Veloz et al., 2020). Many of them are known for stimulating plant defenses, a property which may be officially recognized in the registered usage, as for many of the substances listed in Table 2. A combination of both properties is not rare, and the protective effect of natural substances may also result from additional modes of action. One good example is chitin and chitin-based polymers of D-glucosamine from crab shells, such as chitosan, which have long been known for their ability to elicit plant defenses against plant pathogens, including on horticultural crops (Malerba and Cerana, 2019).

© Burleigh Dodds Science Publishing Limited, 2022. All rights reserved.

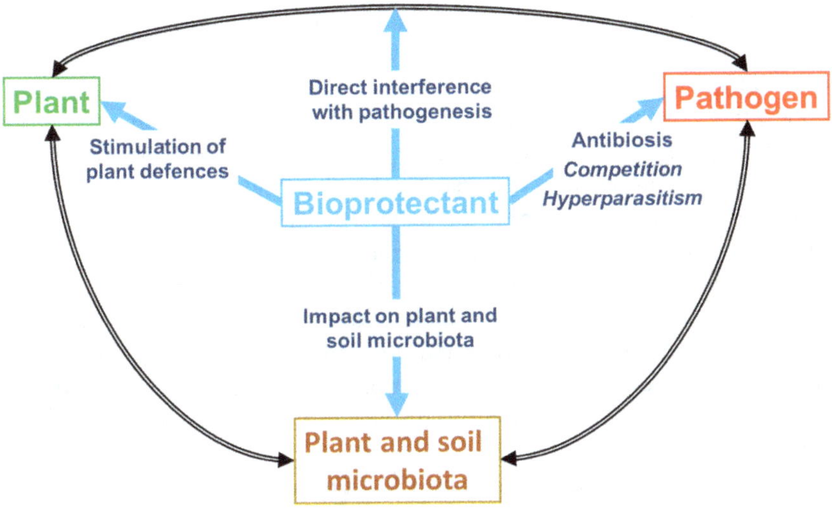

Figure 3 Possible modes of action of bioprotectants used against plant pathogens, either with direct effects, or through an impact on the plant or on its microbiota and that of the soil. Italics indicate those modes of action that are specific to living microorganisms.

In addition, chitosan may provide plant protection through two other possible effects: (i) antimicrobial activity and (ii) film formation on the treated surface, leading to the constitution of a physical barrier against pathogen penetration (Romanazzi et al., 2019). In Europe, it was the first basic substance approved by the European Union for plant protection (Reg. EU 2014/563), for both organic agriculture and integrated pest management (IPM).

In the last decade, spectacular progress has been made in the elucidation of complex mechanisms implicated in the modes of actions of an increasing number of bioprotectants (Köhl et al., 2019b; Sharma et al., 2017). However, one remaining question for those bioprotectants with the capacity to mobilize multiple modes of action, is whether they are all systematically implicated in the protective effect on the plants. One of the rare reports on this issue proposes that defence stimulation represents 30–40% of the activity of chitosan, antibiosis 35–45%, and film-forming activity 20–30%, in terms of its effectiveness in the control of postharvest decay of fresh fruit (Romanazzi et al., 2019). One key challenge for improving field efficacy of bioprotectants will be to identify factors that determine which modes of action are mobilized, singly or in combination and under which conditions.

4.2 Identifying key factors of efficacy

The term 'factor of efficacy' refers to factors that may arise before, during or after the application of a bioprotectant in the field, which could have a positive

© Burleigh Dodds Science Publishing Limited, 2022. All rights reserved.

or a negative impact on the extent of protection it provides to the crop. The factors of efficacy can affect differently the roles of a bioprotectant in reducing the incidence of the targeted disease, avoiding damage to the plant and securing the yield of the crop. Environmental conditions encountered in the field, farming practices, ecology of the targeted plant pathogen, epidemiology of the disease it causes, and biological or biochemical properties and quality of the bioprotectant are among the factors that may impact the efficacy of protection (Aguilar-Veloz et al., 2020; Bardin and Pugliese, 2020). Examples of the main factors of efficacy and their interactions are presented in Fig. 4.

There is abundant scientific literature describing new microbial strains and new compounds with bioprotectant potential against plant diseases, or evaluating their protective efficacy in controlled conditions. However, detailed information on factors governing the efficacy of bioprotectants in the field is still scarce.

Microclimatic conditions (temperature and relative humidity) in the canopy, pH of the targeted area, and quality and intensity of light (ranges of wavelength, photoperiod) have been known for years for their possible impacts on the efficacy of chemical pesticides (Cress, 2006). For bioprotectants, the same main environmental factors can be considered (Abbey et al., 2019). Temperature is probably the most important factor, and one of the most studied. It directly puts to the test the environmental resilience of microbial bioprotectants and can strongly influence their efficacy by affecting their growth rate (Wei et al., 2017), their survival (Landa et al., 2004) and their mechanism of action (Humair et al.,

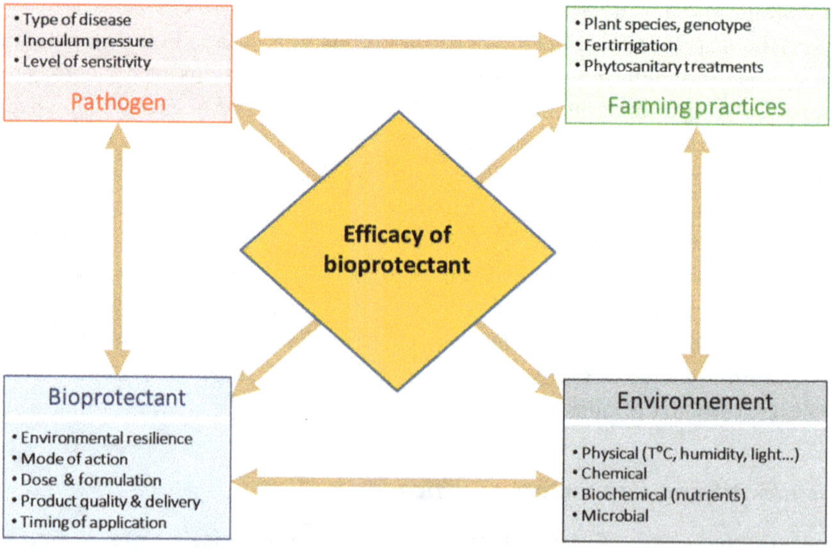

Figure 4 Examples of factors of efficacy of bioprotectants and their possible interactions.

© Burleigh Dodds Science Publishing Limited, 2022. All rights reserved.

2009). For instance, the production of antibiotic compounds and hydrogen cyanide by the root-colonizing bacterial strain *Pseudomonas fluorescens* CHA0 was significantly lower at 35°C than at 30°C (Humair et al., 2009). Temperature also interacts with the effect of other factors such as water activity or relative air humidity, making it important to determine ranges that are optimal, permissive or deleterious for protective efficacy. For example, it was shown for six commercialized bioprotectants that the effect against *B. cinerea* was reduced by 60% if relative humidity decreased below 90%. The optimal temperatures were between 22°C and 25°C, with a strong loss of efficacy below 20°C (Fedele et al., 2020). Environmental conditions are also known to play a major role in the efficacy of natural substances (Aguilar-Veloz et al., 2020). They strongly affect the development of the pathogen (Roger et al., 1999; Elad and Pertot, 2014) and its virulence (Hardwick, 2002), which can ultimately impact the efficacy of a bioprotectant.

The timing of application of the bioprotectants may be critical for successful protection, and can be guided by knowledge of the properties of the bioprotectant, but also of the biology of the pathogen and the epidemiology of the disease. For example, the widely used mycoparasite *Paraphaeosphaeria minitans* (formerly *Coniothyrium minitans*) was found to have a greater effect against sclerotia of *Sclerotinia sclerotiorum* in soil samples collected from field plots treated immediately after a disease outbreak, than from plots treated at a later stage (Gerlagh et al., 2003). A possible explanation for this outcome was that mycoparasitism may be easier on freshly produced sclerotia than on more mature ones. Optimized field application schedules for microbial bioprotectants such as *Bacillus cereus* (SC-1 and P-1) and *Bacillus subtilis* (W-67) against infections by ascospores of *S. sclerotiorum* were focused on the flowering stage of the crop, although they targeted stem rot (Kamal et al., 2015). Appropriate timing for the application of natural substances is also essential (Abbey et al., 2019). For all bioprotectants, their optimization requires knowledge about the persistence of the product on its target, and needs to take into account the microclimatic conditions encountered in the field.

Various factors linked to cultural practices have been studied. For instance, the effect of bioprotectants may be variety-dependent, as was observed for the protection of tomato against *B. cinerea* with resistance-inducing strains of *Trichoderma* (Tucci et al., 2011) or against the seed pathogen *Pythium torulosum* with *Bacillus cereus* UW85 (Smith et al., 1999). Tomato genotypes also had distinct but overlapping root bacterial microbiomes and responded differently to specific bacterial endophytes (French et al., 2020). Plant fertilization may affect the efficacy of bioprotectants. Nitrogen nutrition of the plant was shown to affect the protective effects against *B. cinerea* of two fungal bioprotectants applied to tomato, and of two resistance-inducing products (one based on

© Burleigh Dodds Science Publishing Limited, 2022. All rights reserved.

strain QST713 of *B. subtilis* and one on chitosan) applied to strawberry (Nicot et al., 2013; Abro et al., 2014).

Even though the complexity of factors affecting the efficacy of bioprotectants is often suggested in the scientific literature, most studies only focus on one or two factors at a time or hardly consider what is actually happening in the field. For instance, the effect of temperature, or that of relative humidity, is generally investigated in controlled conditions as static factors (e.g., a constant temperature throughout the experiment) and not as fluctuating factors (e.g., frequent and potentially significant modification during 24 h). This limitation is shared by a large proportion of the publications on bioprotectants, causing possible discrepancies between promising results obtained under controlled conditions in the laboratory, real efficacy observed in the field and feedback from farmers (Ojiambo and Scherm, 2006). Identifying the key factors of efficacy and their effects, alone or in interaction, is an essential step in ensuring the integration of bioprotectants into complex farming systems and thus improving their reliability.

4.3 Improving delivery

Delivery is a critical phase for the protective effect of a bioprotectant, and it can be a major obstacle to its commercial success (Glare et al., 2012). Advances on several issues will be considered here, as delivery is not limited to the context of field application of a bioprotectant by the farmer as presented in section 4.2, but also entails industrial steps such as mass production, formulation, packaging and organization of the supply chain of the product before it reaches the farmers. Throughout this series of steps, one key issue is the maintenance of product stability and preservation of optimal properties of the bioprotectant, to ensure that it will have the capacity to deliver expected bioactive functions when used by the farmer. Other issues are to facilitate effective application and to enhance post-application persistence of the active substance in its target niche. For foliar applications, it was estimated that achieving persistence for more than 3 weeks would constitute a difficult but key goal (Glare et al., 2012). Specific technological details on production and formulation are usually kept as trade secrets by companies, but increasing numbers of publications provide a glimpse into some of the major issues and advances.

At the production stage, optimized technology is needed to ensure that sufficient amounts of bioactive substances are delivered in a cost-effective manner. For microbial bioprotectants, the type of fermentation, its conditions and the composition of the medium are critical. They may also determine the types of propagules that are produced, which in turn influence the stability and the effectiveness of the final product (Zaki et al., 2020). Production conditions may also be manipulated to increase the environmental resilience of the

© Burleigh Dodds Science Publishing Limited, 2022. All rights reserved.

microbial protectant after its application to the crop (Cabrefiga et al., 2011; Daranas et al., 2018). For natural substances such as essential oils and other plant extracts, a high level of standardization is also needed in commercial production. Key factors that may cause variability in the quality of the product were reviewed by Aguilar-Veloz et al. (2020). They include factors pertaining to the plant material, such as the variety and cultivation conditions influenced by climatic factors (photoperiod, shade, temperature, etc.); the plant age (maturity); and agronomic practices (soils, time of planting and time of harvest, application of fertilizers, irrigation, etc.). The type of plant part used in the biological evaluation may also be influential. Other key factors pertain to the experimental procedures used for extraction and conservation and to the methods of chemical characterization (Aguilar-Veloz et al., 2020).

Much progress has been made in optimizing liquid as well as dry formulations, to foster the shelf life and the environmental resilience of microbial bioprotectants (Carbó et al., 2018; Zaki et al., 2020; De La Cruz Quiroz et al., 2019). Inclusion of specific ingredients may play a key role (Marian and Shimizu, 2019). For example, the survival and efficacy of liquid formulations of a strain of *Candida sake* was improved by adding a commercial fatty acid-based product called Fungicover (Carbó et al., 2018). The identification of useful additives may benefit from nutrient profiling of the microbial strains (Gramisci et al., 2018). Further improvements, both for liquid and for dry formulations, should be facilitated by the development of rapid screening tools (combining temperature, water activity and pH conditions) to optimize the formulation of microorganisms for environmental resilience and to predict their ecological niches (Carbó et al., 2018).

Coating agents are increasingly included in the formulations of microbial bioprotectants and natural substances to be used on post-harvest fruits (Marín et al., 2017; Barragán-Menéndez et al., 2020; Aguilar-Veloz et al., 2020). They enhance the protective effect of the active substance, while also providing physiological benefits for the preservation of the produce. For example, the incorporation of the yeast *Meyerozyma caribbica* in a sodium alginate matrix greatly increased its protective effect against anthracnose caused by *Colletotrichum gloeosporioides* on avocado fruits, reaching 100% protection at 6°C, and it reduced weight loss due to dehydration (Iñiguez-Moreno et al., 2020). Lactic acid bacteria such as *Lactobacillus plantarum*, incorporated into various edible coatings, were shown to provide control of decay fungi on grape berries and to maintain post-harvest quality (Marín et al., 2019). Besides microbial bioprotectants, coatings with a range of molecules and natural products incorporated into chitosan-based polymers were shown to provide protection of blueberry, jujube and strawberry fruits against post-harvest diseases (Malerba and Cerana, 2019).

Other reported additives for the formulation of microbial bioprotectants include dextrin, milk, molasses, oils, pectin, starch and sucrose (Zaki et al.,

© Burleigh Dodds Science Publishing Limited, 2022. All rights reserved.

2020; De La Cruz Quiroz et al., 2019). Some of these compounds contribute to the microencapsulation of the bioprotectants with biopolymers, a technology for which tremendous advances have been made in recent years. Microencapsulation consists of creating a physical barrier that provides protection of the active ingredient against external factors such as mechanical stress, ultraviolet radiation, oxidation and high temperatures. As a result, the microbial bioprotectants are able to survive for a longer period and their metabolic activity is maintained (Fraceto et al., 2018). Advances in microencapsulation have also been significant for the formulation of natural substances such as plant extracts, as reviewed by Aguilar-Veloz et al. (2020). Improved formulations have contributed to better control of post-harvest rot of horticultural produce, as well as reducing the mycotoxin contents. Enhanced control of post-harvest rot of sweet potato, caused by fungal pathogens such as *Rhizopus stolonifer* and *Ceratocystis fimbriata*, was recently achieved using the lipopeptide iturin A with a new microencapsulation technology based on chitosan (Yuan et al., 2020). After 50 days of storage, the incidence of rot was 8% in tubers treated with the micro-encapsulated iturin, compared to 14% with non-encapsulated iturin, and nearly 56% for the untreated control. It also significantly decreased the tuber content of ipomeamarone, a toxic furanoterpenoid which accumulates in response to microbial attack in rotting sweet potatoes (Wamalwa et al., 2015). The content of this compound was 2 mg/100g in tubers treated with micro-encapsulated iturin, compared to 19 mg/100 g for tubers treated with the non-encapsulated iturin and 135 mg/100 g for the untreated control (Yuan et al., 2020).

There is increasing interest in formulations based on nanotechnologies both for microbial bioprotectants (Fraceto et al., 2018) and for natural substances (Aguilar-Veloz et al., 2020). Recent application of chitosan-based polymers and nanoparticles complexed with a range of molecules and natural substances were shown to increase the resistance of horticultural crops such as tomato, pepper and lettuce against various soil-borne pathogens (Malerba and Cerana, 2019). However, further research will be necessary to optimize and confirm the potential of these technologies, and to support decision-makers in the development of adapted regulations and safety assessment procedures (Aguilar-Veloz et al., 2020; Fraceto et al., 2018).

Advances in packaging technology may also contribute to the stability of bioprotectants and the extension of their shelf life, by controlling gas exchange, preventing the loss or gain of moisture and avoiding contamination of the product (Gotor-Vila et al., 2019). Successful packaging optimization was recently reported for formulated preparations of the strain CPA-8 of *Bacillus amyloliquefaciens*, with significant positive effects for long-term storability at 22°C while providing suitable product delivery and field application, as well as the effective control of *Monilinia* brown rot of nectarines (Gotor-Vila et al., 2019).

© Burleigh Dodds Science Publishing Limited, 2022. All rights reserved.

4.4 Complementing bioprotectant use with other solutions from the integrated pest management toolbox

IPM approaches have long been investigated for the management of plant health in horticultural crops. As early as 1959, a 'Working Group for Integrated Plant Protection in Fruit Orchards' was established by the International Organization for Biological and Integrated Control of Noxious Animals and Plants (IOBC), followed by working groups on integrated protection for greenhouse crops in 1968 and vegetable crops in 1970 (Freier and Boller, 2009; Boller et al., 2006). While early work on IPM was mostly focused on pest animals (particularly invertebrates), the concept was gradually extended to include the control of plant diseases, and the word 'pest' is increasingly understood to include weeds and causal agents of diseases as well as pest animals (van Lenteren and Nicot, 2020). Bioprotectants are particularly well suited for use in IPM programs and their application can be combined with a range of other plant protection practices that can enhance their efficacy (van Lenteren and Nicot, 2020; Glare et al., 2012).

The bioprotection-enhancing effect of agricultural practices may result from a reduction of disease pressure on the crop, for example by avoiding conditions that foster severe forms of disease expression or epidemic development, or by decreasing plant susceptibility (through the planting of less susceptible varieties or through interventions that stimulate plant defenses). This may be achieved by combining the application of bioprotectants with abiotic or biotic stresses. The application of different abiotic and biotic stresses on the roots of tomato plants, for example, was shown to enhance chemotropism and the growth of a biocontrol strain of *T. harzianum* in the rhizosphere (Lombardi et al., 2018). Fertilization may be a particularly interesting lever. For example, appropriate levels of nitrogen fertilization have a potential to reduce the susceptibility of various horticultural crops to *B. cinerea* and *S. sclerotiorum*, but also to increase the protective effect of bioprotectants (Nicot et al., 2013; Lecompte et al., 2010; Abro et al., 2014; Abro et al., 2013; Lecompte et al., 2013).

A large body of literature describes possible bioprotection-enhancing effects of soil organic amendments, such as composts, manures and organic wastes on the efficacy of microbial bioprotectants (Marian and Shimizu, 2019). Composts amended with strains of microbial bioprotectants have shown high potential for disease control. Recent examples include the efficient reduction in the incidence and severity of Fusarium wilt of lettuce, provided by incorporating into the soil a compost-based biofertilizer amended with the strain TW2 of *Trichoderma* sp. (Gilardi et al., 2019), and the highly efficient control of *Pythium* infection in tomato by strain 4.4.1 of *Pseudomonas chlororaphis*, and various organic amendments (Postma and Nijhuis, 2019). Organic amendments may act as nutrients that are preferentially used by specific microbial bioprotectants, but there is also increasing interest in their global effects on indigenous beneficial microorganisms present

© Burleigh Dodds Science Publishing Limited, 2022. All rights reserved.

in the soil, to foster soil suppressiveness against soil-borne plant pathogens (Tao et al., 2020; De Corato, 2020; Mazzola and Freilich, 2017).

Enhanced disease control may also be achieved above ground by combining natural substances with microbial bioprotectants or by combining several microbial strains with synergistic or complementary capacities. Chitosan has been combined successfully with various microorganisms for the post-harvest protection of horticultural produce (Zhou et al., 2016; Godana et al., 2020). For example, enhanced protection of grape berries against *Penicillium expansum* was achieved by the addition of 1% chitosan to preparations of biocontrol yeast *Pichia anomala* (Godana et al., 2020). The increased protection was associated with increased pullulation of yeast cells in the wounds in the presence of chitosan. The combined application of chitosan with extracts from the brown alga *Ascophyllum nodosum* provided significantly reduced development of pea powdery mildew by a synergistic activation of the plant's defenses (Patel et al., 2020). Several strains of microbial bioprotectants are already commercialized in combinations (Table 1). Combinations of microbial strains belonging to the same species or to species of the same genus are already commercialized for the protection of horticultural crops (Table 1). More may come in the future, with increasing numbers of reports of successful multispecies assemblages as reviewed by Marian and Shimizu (2019), such as mixtures of Bacillus, Pseudomonas and Rhizobium that were shown to provide yield improvement of bean together with control of Fusarium wilt (Kalantari et al., 2018). Ultimately, microbiome research may lead to the screening or engineering of more complex consortia of microbial bioprotectants (Kehe et al., 2019; Mitter et al., 2019; Woo and Pepe, 2018).

5 Fostering the adoption of biocontrol

The reticence of farmers to use bioprotectants has long been considered to be an obstacle to the wide adoption of bioprotectants for disease management, both in horticultural and other production systems (Barratt et al., 2018; Lamichhane et al., 2017; Abbey et al., 2019; Villemaine et al., 2020; Marrone, 2019). This section will present an example of current use of bioprotectants by horticulturists in France, considerations on possible obstacles for wider adoption, and outlooks for helping farmers cope with the technical complexity of deploying bioprotectants in IPM strategies.

5.1 Case study: adoption of bioprotectants by horticulturists in France

To gain a better perspective on bioprotectant use in horticultural crops, we recently carried out a survey of farmers, farm advisors and farm suppliers. A

© Burleigh Dodds Science Publishing Limited, 2022. All rights reserved.

questionnaire was widely disseminated through farmers' associations, advisory services and in technical journals. Among 170 responses, 90 were received from farm advisors, 53 from horticulturists, (essentially fruit and vegetable farmers) and 27 from suppliers of bioprotectants. Among responding horticulturists, 41% (22 of the 53) declared that they used one or more microbial bioprotectants for disease management. This figure may overestimate the frequency of use of microbial bioprotectants, as the survey was clearly focused on bioprotection and non-users may have chosen not to respond. Nearly half of the responding horticulturists (26 of the 53) had some organic production, and among them 16 used both microbial bioprotectants and natural products for disease control, while 10 used only natural products. Microbial bioprotectants were used by 6 of the 27 conventional farmers. The survey generated a greater sample size for farm advisors. Among the 90 respondents, 77 declared that they recommended microbial bioprotectants for disease control. Again, this high number may be biased by the fact that the questionnaire was focused on bioprotection. Microbial bioprotectants were most often recommended for the control of *Sclerotinia* white mold, powdery mildews and *Botrytis* gray mold, with reported recommendations by 32–39% of the farm advisors (Fig. 5). The recommendation of microbial bioprotectants appeared to be somewhat limited for the control of bacterial diseases, and viral diseases were the least targeted, with recommendations of bioprotectant use by only 4 % of the farm advisors.

Twenty microbial bioprotectants commercialized in France were recommended by the farm advisors for disease management of horticultural crops (Fig. 6). They included 13 of the 16 microbial preparations based on fungal strains shown in Fig. 1: 6 Trichoderma-based products (labeled A, C, E, H, M and P in Fig. 6); 4 yeasts (labeled B, K, N and O) and 3 other fungal preparations (labeled D, Q and S). The

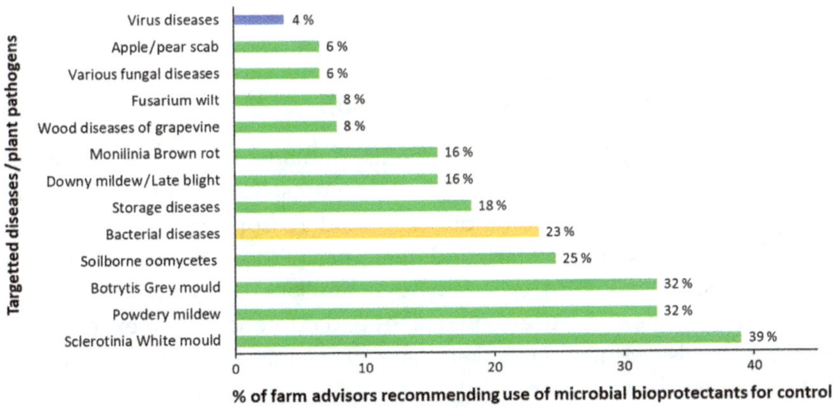

Figure 5 Key diseases and plant pathogens targeted by recommendations of microbial bioprotectant use made by 77 French farm advisors. Blue, green and yellow bars indicate viral, fungal and bacterial targets, respectively.

© Burleigh Dodds Science Publishing Limited, 2022. All rights reserved.

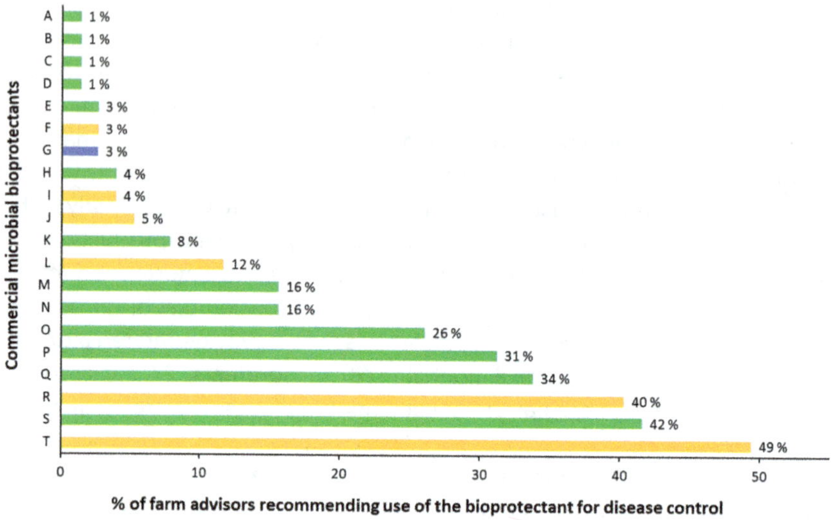

Figure 6 Frequency of recommendation of 20 commercial microbial bioprotectants (anonymously labeled from A to T) for disease control of horticultural crops by 77 French farm advisors. The colors of the bars indicate the type of microorganisms present as active substance in the bioprotectants: green = fungus; yellow = bacterium; blue = virus.

six bacterium-based bioprotectants prescribed by the farm advisors (out of eight shown in Fig. 1) included five *Bacillus*-based preparations (labeled F, J, L, R and T in Fig. 6) and one other bacterium (labeled I).

Although some products are recommended by nearly half of the farm advisors, there is clearly room for wider adoption of microbial bioprotectants, as 5 products were not prescribed at all and 11 were recommended by fewer than 10% of the farm advisors.

5.2 Possible obstacles for wider adoption

In the survey mentioned, certain respondents spontaneously offered comments to express their reticence about the use or the prescription of bioprotectants. Among farmers, the most commonly reported reason was the absence of knowledge about the products and their use. Products were sometimes unavailable locally. Among farm advisors, the lack of information on the products and the need for users' guidelines and references from experimental stations were also frequently evoked. Lack of reliable field efficacy and high costs were also given as reasons.

These survey returns corroborate, for French horticulturists, previously reported analyses of possible impediments to the adoption of bioprotectants

© Burleigh Dodds Science Publishing Limited, 2022. All rights reserved.

at a more global scale (Marrone, 2009, 2019; Moser et al., 2008; Abdollahzadeh et al., 2017; Lamichhane et al., 2017). Marrone (2019) recently described several challenges to the adoption of biopesticides: (i) lack of awareness and education on how to deploy their unique modes of action in integrated programs, (ii) testing products alone versus in integrated programs and (iii) lingering perceptions of high cost and low efficacy.

Socio-economic factors may also play a role, as well as possible issues with risk aversion and acceptance of novelty, as shown from a survey-based study on the prescription of bioprotectants for wine-grapes in the Burgundy region of France (Villemaine et al., 2020) and other studies (Moser et al., 2008; Abdollahzadeh et al., 2017).

The limited number of bioprotectants currently available for disease control, together with the general situation with minor uses presented in part 2 of this chapter, may be complicating factors. In the absence of available bioprotectants for the control of a key disease, a farmer will be likely to resort to chemical control. This, in turn, may interfere with the activity of other microbial bioprotectants and reduce their efficacy. This hypothetical situation highlights the need to provide farmers with detailed information on the compatibility of bioprotectants with other plant protection products.

One key stumbling block for widening the adoption of bioprotectants by farmers may be their perceived or actually encountered lack of reliable efficacy in the field (Bardin and Pugliese, 2020; Abbey et al., 2019). Bioprotectants, and especially microbials, are generally more sensitive than synthetic chemicals to environmental factors (Bailey et al., 2010; Marrone, 2009). They are also much more complex to use, in relation to the intricate modes of actions and factors of efficacy described in this chapter. Comprehending this complexity is a major challenge for the research community, as it requires advanced multidisciplinary expertise in biology, physics and (bio)chemistry, among others. For the farmers, harnessing this complexity to make optimal decisions for the deployment of bioprotectants is, understandably, also a tremendous challenge. This is compounded by the fact that many farmers have ample experience with the application of synthetic pesticides, but no or very limited experience with the use of bioprotectants.

5.3 Coping with the technical complexity of deployment

One possible way to help farmers overcome this challenge is the development of decision support systems (DSS) that would translate scientific knowledge on the biology and modes of action of bioprotectants into simple and hierarchized decision trees. In relation to the digitalization of agriculture, DSS are becoming increasingly available to farmers, for example for the optimization of fertilization and irrigation (Zhai et al., 2020). Excellent reviews have highlighted the

© Burleigh Dodds Science Publishing Limited, 2022. All rights reserved.

potential (and complexity) of DSS for IPM, based on models that forecast the risk of disease development (Damos, 2015; Rossi et al., 2019; Guimapi et al., 2020). To date however, there is a shortage of tools to support decisions on the deployment of bioprotectants for disease control. To our knowledge, only two have been reported. One is 'BOTMAN', a weather-based tool designed to assist decisions on the utilization of a Trichoderma bioprotectant, or alternatively chemical pesticides, for the control of *B. cinerea* in tomato and cucumber greenhouses (Shtienberg and Elad, 1997). More recently, a weather-based disease forecasting model, designed to help farmers schedule copper sprays against Cercospora spot of avocado, was extended to include elements about the behavior of a microbial bioprotectant (Van Eeden and Korsten, 2003).

Work has been initiated in our group to develop knowledge-driven DSS to facilitate the deployment of microbial bioprotectants against diseases of horticultural crops. The first objective is to identify and prioritize the factors governing the efficacy of microbial biocontrol agents and then to identify those that could be modulated by farmers' practices. As a second step, a generic DSS prototype will be designed to integrate key factors of efficacy and their interactions with each other. The final step will be to parameterize and evaluate a DSS for a specific crop-bioprotectant-pathogen combination.

6 Conclusion

Microbial bioprotectants and natural substances have been increasingly available and deployed for disease management, contributing to enhanced sustainability of horticultural production. Because of the large number of horticultural plant species, and of pathogens that can affect them, currently registered products cover only a fraction of farmers' needs for disease control solutions. However, biocontrol and IPM constitute a highly dynamic field of research, both in academia and in the plant protection industry, and high numbers of new bioprotectants are likely to be developed in the coming years, covering an increasing proportion of the needs. The exploration of a broader range of possible sources can be expected to provide a much wider diversity of microbial species and natural substances. Meanwhile, expanding knowledge about the properties of bioprotectants currently on the market should result in significant widening of their spectrum of action, to cover more disease targets on more horticultural plant species. Even entomopathogenic microorganisms traditionally used against pest insects may prove to be useful for disease control (Sarven et al., 2020).

7 Future trends in research

One key challenge to further increase adoption of bioprotectants by farmers is to enhance and stabilize their field efficacy. Continued improvements in

© Burleigh Dodds Science Publishing Limited, 2022. All rights reserved.

their formulation can play a determinant role, for example in sheltering the bioprotectants from detrimental environmental conditions. Novel formulations may also increasingly foster their bioactivity, with increased knowledge about the modes of action of the bioprotectants. This knowledge is also likely to support the design of bioprotectants combining microbial strains or natural substances with complementary or synergistic modes of action. One promising avenue derived from microbiome research is the engineering of microbial consortia to support bioprotectants that are applied, or with their own protective effect (Eckshtain-Levi et al., 2020; Ab Rahman et al., 2018). This research is also likely to lead to innovative tools to recruit indigenous rhizosphere microbiomes to develop suppressiveness against soil-borne diseases (Schlatter et al., 2017; Mazzola and Freilich, 2017).

Improved field efficacy will also result from the inclusion of bioprotectants in IPM strategies. Bioprotectants for plant disease control have long been regarded as substitution tools, to be applied as sustainable alternatives to synthetic pesticides. This concept is conveyed by the word 'biopesticide', which is commonly used to designate certain bioprotectants. Although bioprotectants may often have the ability to provide full protection of crops on their own (van Lenteren et al., 2018), they are decreasingly considered to be stand-alone solutions, as has been expected with chemical pesticides. Rather, bioprotectants are increasingly regarded as components of a toolbox of IPM solutions, which can be combined to design strategies for plant health management (Abbey et al., 2019). This is particularly true in the context of horticultural crops, for which IPM is already a quite common practice and which benefits from a wide range of possible plant health management tools.

Finally, a key challenge for the future will be to enable farmers to cope with the complexity attached to achieving optimized integration of bioprotectants in IPM schemes. For this, it will be necessary to acquire and translate detailed and multifaceted knowledge into simple decision rules that can be implemented at field level, and to deliver them within user-friendly DSS.

8 Where to look for further information

The following articles provide a good overview of key aspects of the subject:

- Ab Rahman, S. F. S., Singh, E., Pieterse, C. M. J. and Schenk, P. M. (2018), Emerging microbial biocontrol strategies for plant pathogens. *Plant Science*, 267, 102–111.
- Bardin, M. and Pugliese, M. (2020), Biocontrol agents against diseases. In: Gullino, M., Albajes , R. and Nicot, P. (Eds) *Integrated Pest and Disease Management in Greenhouse Crops*. Springer.

© Burleigh Dodds Science Publishing Limited, 2022. All rights reserved.

- Barratt, B., Moran, V., Bigler, F. and van Lenteren, J. (2018), The status of biological control and recommendations for improving uptake for the future. *BioControl*, 63, 155–167.
- Compant, S., Samad, A., Faist, H. and Sessitsch, A. (2019), A review on the plant microbiome: Ecology, functions, and emerging trends in microbial application. *Journal of Advanced Research*, 19, 29–37.
- De Silva, N. I., Brooks, S., Lumyong, S. and Hyde, K. D. (2019), Use of endophytes as biocontrol agents. *Fungal Biology Reviews*, 33, 133–148.
- Köhl, J., Kolnaar, R. and Ravensberg, W. J. (2019), Mode of action of microbial biological control agents against plant diseases: relevance beyond efficacy. *Frontiers in Plant Science*, 10, 845. doi: 10.3389/fpls.2019.00845.
- Raveau, R., Fontaine, J. and Sahraoui, A. L. H. (2020), Essential oils as potential alternative biocontrol products against plant pathogens and weeds: A Review. *Foods,* 9, 31.
- Raymaekers, K., Ponet, L., Holtappels, D., Berckmans, B. and Cammue, B. P. A. (2020), Screening for novel biocontrol agents applicable in plant disease management – A review. *Biological Control*, 144, 104240.
- Zhai, Z., Martínez, J. F., Beltran, V. and Martínez, N. L. (2020), Decision support systems for agriculture 4.0: Survey and challenges. *Computers and Electronics in Agriculture,* 170, 105256.

Key research in this area can be found at member institutions of the International Organisation of Biological Control (https://www.iobc-global.org/) and those of scientists who participate in the activities of Working Group of its West Palaearctic Regional Section on the following topics:

- Biological and integrated control of Plant Pathogens (https://www.iobc-wprs.org/expert_groups/06_wg_plant_pathogens.html).
- Induced resistance in plants against insects and diseases (https://www.iobc-wprs.org/expert_groups/17_wg_induced_resistance.html).
- Integrated control in Protected crops, Temperate climate (https://www.iobc-wprs.org/expert_groups/08_wg_protected_crops_temperate.html) and Mediterranean climate (https://www.iobc-wprs.org/expert_groups/09_wg_protected_crops_mediterranean.html).
- Integrated control in Citrus fruit crops (https://www.iobc-wprs.org/expert_groups/16_wg_citrus_crops.html).
- Integrated protection of Field vegetable Crops (https://www.iobc-wprs.org/expert_groups/12_wg_vegetables.html).
- Integrated protection of Fruit crops (https://www.iobc-wprs.org/expert_groups/02_wg_fruit_crops.html).
- Integrated protection of Olive crops (https://www.iobc-wprs.org/expert_groups/15_wg_olive_crops.html).

© Burleigh Dodds Science Publishing Limited, 2022. All rights reserved.

9 References

Ab Rahman, S. F. S., Singh, E., Pieterse, C. M. J. and Schenk, P. M. 2018. Emerging microbial biocontrol strategies for plant pathogens. *Plant Science: An International Journal of Experimental Plant Biology* 267, 102-111.

Abbey, J. A., Percival, D., Abbey, L., Asiedu, S. K., Prithiviraj, B. and Schilder, A. 2019. Biofungicides as alternative to synthetic fungicide control of grey mould (*Botrytis cinerea*)–prospects and challenges. *Biocontrol Science and Technology* 29(3), 207-228.

Abdollahzadeh, G., Damalas, C. A. and Sharifzadeh, M. S. 2017. Understanding adoption, non-adoption, and discontinuance of biological control in rice fields of northern Iran. *Crop Protection* 93, 60-68.

Abro, M. A., Lecompte, F., Bryone, F. and Nicot, P. C. 2013. Nitrogen fertilization of the host plant influences production and pathogenicity of *Botrytis cinerea* secondary inoculum. *Phytopathology* 103(3), 261-267.

Abro, M. A., Lecompte, F., Bardin, M. and Nicot, P. C. 2014. Nitrogen fertilization impacts biocontrol of tomato gray *Agronomy for Sustainable Development* 34(3), 641-648.

Aguilar-Veloz, L. M., Calderón-Santoyo, M., Vázquez González, Y. and Ragazzo-Sánchez, J. A. 2020. Application of essential oils and polyphenols as natural antimicrobial agents in postharvest treatments: advances and challenges. *Food Science and Nutrition* 8(6), 2555-2568.

Antoniou, A., Tsolakidou, M. D., Stringlis, I. A. and Pantelides, I. S. 2017. Rhizosphere microbiome recruited from a suppressive compost improves plant fitness and increases protection against vascular wilt pathogens of tomato. *Frontiers in Plant Science* 8, 2022.

Arrarte, E., Garmendia, G., Rossini, C., Wisniewski, M. and Vero, S. 2017. Volatile organic compounds produced by Antarctic strains of *Candida sake* play a role in the control of postharvest pathogens of apples. *Biological Control* 109, 14-20.

Bailey, K. L., Boyetchko, S. M. and Längle, T. 2010. Social and economic drivers shaping the future of biological control: a Canadian perspective on the factors affecting the development and use of microbial biopesticides. *Biological Control* 52(3), 221-229.

Bardin, M. and Pugliese, M. 2020. Biocontrol agents against diseases. In: Gullino, M., Albajes , R. and Nicot, P. (Eds) *Integrated Pest and Disease Management in Greenhouse Crops*, pp. 385-407. Springer. Available at: https://www.springer.com/gp/book/9783030223038.

Barragán-Menéndez, C., Gálvez-López, D., Rosas-Quijano, R., Salvador-Figueroa, M., Ovando-Medina, I. and Vázquez-Ovando, A. 2020. Films of chitosan and aloe vera for maintaining the viability and antifungal activity of *Lactobacillus paracasei* TEP6. *Coatings* 10(3), 259. https://doi.org/10.3390/coatings10030259

Barratt, B. I. P., Moran, V. C., Bigler, F. and van Lenteren, J. C. 2018. The status of biological control and recommendations for improving uptake for the future. *BioControl* 63(1), 155-167.

Benhamou, N. 2004. Potential of the mycoparasite, *Verticillium lecanii*, to protect citrus fruit against *Penicillium digitatum*, the causal agent of green mold: a comparison with the effect of chitosan. *Phytopathology* 94(7), 693-705.

Berg, G., Koberl, M., Rybakova, D., Muller, H., Grosch, R. and Smalla, K. 2017. Plant microbial diversity is suggested as the key to future biocontrol and health trends. *FEMS Microbiology Ecology* 93(5), 9.

© Burleigh Dodds Science Publishing Limited, 2022. All rights reserved.

Besset-Manzoni, Y., Joly, P., Brutel, A., Gerin, F., Soudiere, O., Langin, T. and Prigent-Combaret, C. 2019. Does in vitro selection of biocontrol agents guarantee success in planta? A study case of wheat protection against Fusarium seedling blight by soil bacteria. *PLoS ONE* 14(12), e0225655.

Boller, E. F., van Lenteren, J. C. and Delucchi, V. (Eds) 2006. International Organization for Biological Control of Noxious Animals and Plants (IOBC): History of the first 50 years (1956–2006). Available at: http://www.iobc-wprs.org/members/shop_en.cfm?mod_Shop_detail_produkte=210.

Bouaoud, Y., Troulet, C., Foughalia, A., Berge, O., Aissat, K. and Bardin, M. 2018. A multi-criteria approach for the selection of efficient biocontrol agents against *Botrytis cinerea* on tomato in Algeria. *BioControl* 63(2), 299–311.

Cabrefiga, J., Francés, J., Montesinos, E. and Bonaterra, A. 2011. Improvement of fitness and efficacy of a fire blight biocontrol agent via nutritional enhancement combined with osmoadaptation. *Applied and Environmental Microbiology* 77(10), 3174–3181.

Carbó, A., Torres, R., Teixidó, N., Usall, J., Magan, N. and Medina, A. 2018. Predicted ecological niches and environmental resilience of different formulations of the biocontrol yeast *Candida sake* CPA-1 using the Bioscreen C. *BioControl* 63(6), 855–866.

Chandler, D., Bailey, A. S., Tatchell, G. M., Davidson, G., Greaves, J. and Grant, W. P. 2011. The development, regulation and use of biopesticides for integrated pest management. *Philosophical Transactions of the Royal Society of London. Series B, Biological Sciences* 366(1573), 1987–1998.

Chen, X., Gao, Z., Swisher, M., House, L. and Zhao, X. 2018. Eco-labeling in the fresh produce market: not all environmentally friendly labels are equally valued. *Ecological Economics* 154, 201–210.

Compant, S., Samad, A., Faist, H. and Sessitsch, A. 2019. A review on the plant microbiome: ecology, functions, and emerging trends in microbial application. *Journal of Advanced Research* 19, 29–37.

Conrath, C., Beckers, G. J., Langenbach, C. J. and Jaskiewicz, M. R. 2015. Priming for enhanced defense. *Annual Review of Phytopathology* 53, 97–119.

Coppola, M., Cascone, P., Lelio, I. D., Woo, S. L., Lorito, M., Rao, R., Pennacchio, F., Guerrieri, E. and Digilio, M. C. 2019. *Trichoderma atroviride* P1 colonization of tomato plants enhances both direct and indirect defense barriers against insects. *Frontiers in Physiology* 10, 813.

Cress, D. 2006. Factors affecting pesticide behavior. *Greenhouse Product News*, September issue, 28–33.

Damos, P. 2015. Modular structure of web-based decision support systems for integrated pest management: a review. *Agronomy for Sustainable Development* 35(4), 1347–1372.

Daranas, N., Badosa, E., Francés, J., Montesinos, E. and Bonaterra, A. 2018. Enhancing water stress tolerance improves fitness in biological control strains of *Lactobacillus plantarum* in plant environments. *PLoS ONE* 13(1), e0190931.

De Corato, U. 2020. Disease-suppressive compost enhances natural soil suppressiveness against soil-borne plant pathogens: a critical review. *Rhizosphere* 13, 100192.

De La Cruz Quiroz, R., Cruz Maldonado, J. J., Rostro Alanis, MdJ., Torres, J. A. and Parra Saldívar, R. 2019. Fungi-based biopesticides: shelf-life preservation technologies used in commercial products. *Journal of Pest Science* 92(3), 1003–1015.

© Burleigh Dodds Science Publishing Limited, 2022. All rights reserved.

De Silva, N. I., Brooks, S., Lumyong, S. and Hyde, K. D. 2019. Use of endophytes as biocontrol agents. *Fungal Biology Reviews* 33(2), 133–148.

Donley, N. 2019. The USA lags behind other agricultural nations in banning harmful pesticides. *Environmental Health: A Global Access Science Source* 18(1), 44.

Eckshtain-Levi, N., Harris, S. L., Roscios, R. Q. and Shank, E. A. 2020. Bacterial community members increase bacillus subtilis maintenance on the roots of *Arabidopsis thaliana*. *Phytobiomes Journal* 4(4), 303–313.

Elad, Y. and Pertot, I. 2014. Climate change impacts on plant pathogens and plant diseases. *Journal of Crop Improvement* 28(1), 99–139.

Ellis, J. G. 2017. Can plant microbiome studies lead to effective biocontrol of plant diseases? *Molecular Plant-Microbe Interactions: MPMI* 30(3), 190–193.

Elsayed, T. R., Jacquiod, S., Nour, E. H., Sorensen, S. J. and Smalla, K. 2020. Biocontrol of bacterial wilt disease through complex interaction between tomato plant, antagonists, the indigenous rhizosphere microbiota, and *Ralstonia solanacearum*. *Frontiers in Microbiology* 10, 2835.

FAO and WHO 2017. International Code of Conduct on Pesticide Management: Guidelines for the Registration of Microbial, Botanical and Semiochemical Pest Control Agents for Plant Protection and Public Health Uses. The Food and Agriculture Organization of the United Nations (FAO) and the World Health Organization (WHO).

Fedele, G., Brischetto, C. and Rossi, V. 2020. Biocontrol of *Botrytis cinerea* on grape berries as influenced by temperature and humidity. *Frontiers in Plant Science* 11, 1232.

Fraceto, L. F., Maruyama, C. R., Guilger, M., Mishra, S., Keswani, C., Singh, H. B. and De Lima, R. 2018. Trichoderma harzianum-based novel formulations: potential applications for management of Next-Gen agricultural challenges. *Journal of Chemical Technology and Biotechnology* 93(8), 2056–2063.

Frederiks, C. and Wesseler, J. H. H. 2019. A comparison of the EU and US regulatory frameworks for the active substance registration of microbial biological control agents. *Pest Management Science* 75(1), 87–103.

Freier, B. and Boller, E. F. 2009. Integrated pest management in Europe-history, policy, achievements and implementation. In: Peshin R. and Dhawan A. K. (Eds) *Integrated Pest Management: Dissemination and Impact*, pp. 435–454. Dordrecht: Springer. https://doi.org/10.1007/978-1-4020-8990-9_14.

French, E., Tran, T. and Iyer-Pascuzzi, A. S. 2020. Tomato genotype modulates selection and responses to root microbiota. *Phytobiomes Journal* 4(4), 314–326.

Gao, Z., Zhang, B., Liu, H., Han, J. and Zhang, Y. 2017. Identification of endophytic *Bacillus velezensis* ZSY-1 strain and antifungal activity of its volatile compounds against *Alternaria solani* and *Botrytis cinerea*. *Biological Control* 105, 27–39.

Gerlagh, M., Goossen-Van De Geijn, H. M., Hoogland, A. E. and Vereijken, P. F. G. 2003. Quantitative aspects of infection of *Sclerotinia sclerotiorum* sclerotia by *Coniothyrium minitans*–timing of application, concentration and quality of conidial suspension of the mycoparasite. *European Journal of Plant Pathology* 109(5), 489–502.

Gilardi, G., Pugliese, M., Gullino, M. L. and Garibaldi, A. 2019. Nursery treatments with resistant inducers, soil amendments and biocontrol agents for the management of the Fusarium wilt of lettuce under glasshouse and field conditions. *Journal of Phytopathology* 167(2), 98–110.

© Burleigh Dodds Science Publishing Limited, 2022. All rights reserved.

Glare, T., Caradus, J., Gelernter, W., Jackson, T., Keyhani, N., Köhl, J., Marrone, P., Morin, L. and Stewart, A. 2012. Have biopesticides come of age? *Trends in Biotechnology* 30(5), 250–258.

Godana, E. A., Yang, Q., Wang, K., Zhang, H., Zhang, X., Zhao, L., Abdelhai, M. H. and Guillaume Legrand, N. N. 2020. Bio-control activity of *Pichia anomala* supplemented with chitosan against *Penicillium expansum* in postharvest grapes and its possible inhibition mechanism. *LWT* 124, 109188.

Gotor-Vila, A., Teixido, N., Sisquella, M., Torres, R. and Usall, J. 2017. Biological characterization of the biocontrol agent bacillus amyloliquefaciens CPA-8: the effect of temperature, pH and water activity on growth, susceptibility to antibiotics and detection of enterotoxic genes. *Current Microbiology* 74(9), 1089–1099.

Gotor-Vila, A., Usall, J., Torres, R., Solsona, C. and Teixido, N. 2019. Enhanced shelf-life of the formulated biocontrol agent *Bacillus amyloliquefaciens* CPA-8 combining diverse packaging strategies and storage conditions. *International Journal of Food Microbiology* 290, 205–213.

Gramisci, B. R., Lutz, M. C., Lopes, C. A. and Sangorrín, M. P. 2018. Enhancing the efficacy of yeast biocontrol agents against postharvest pathogens through nutrient profiling and the use of other additives. *Biological Control* 121, 151–158.

Gu, K. B., Zhang, D. J., Guan, C., Xu, J. H., Li, S. L., Shen, G. M., Luo, Y. C. and Li, Y. G. 2017. Safe antifungal lipopeptides derived from *Bacillus marinus* B-9987 against grey mold caused by *Botrytis cinerea*. *Journal of Integrative Agriculture* 16(9), 1999–2008.

Guimapi, R. A., Mohamed, S. A., Biber-Freudenberger, L., Mwangi, W., Ekesi, S., Borgemeister, C. and Tonnang, H. E. Z. 2020. Decision support system for fitting and mapping nonlinear functions with application to insect pest management in the biological control context. *Algorithms* 13, 20.

Guimarães, R. A., Pherez-Perrony, P. E., Müller, H., Berg, G., Medeiros, F. H. V. and Cernava, T. 2020. Microbiome-guided evaluation of *Bacillus subtilis* BIOUFLA2 application to reduce mycotoxins in maize kernels. *Biological Control* 150, 104370.

Hao, L., Wang, Y., Chen, X., Zheng, X., Chen, S., Li, S., Zhang, Y. and Xu, Y. 2019. Exploring the potential of natural products from mangrove rhizosphere bacteria as biopesticides against plant diseases. *Plant Disease* 103(11), 2925–2932.

Hardwick, N. V. 2002. Weather and plant diseases. *Weather* 57(5), 184–190.

Hu, X., Cheng, B., Du, D., Huang, Z., Pu, Z., Chen, G., Peng, A. and Lu, L. 2019. Isolation and identification of a marine actinomycete strain and its control efficacy against citrus green and blue moulds. *Biotechnology and Biotechnological Equipment* 33(1), 719–729.

Hubbard, M., Hynes, R. K., Erlandson, M. and Bailey, K. L. 2014. The biochemistry behind biopesticide efficacy. *Sustainable Chemical Processes* 2(1), 18.

Humair, B., González, N., Mossialos, D., Reimmann, C. and Haas, D. 2009. Temperature-responsive sensing regulates biocontrol factor expression in *Pseudomonas fluorescens* CHA0. *The ISME Journal* 3(8), 955–965.

IBMA 2019. Definition: bioprotection as the global term for all biocontrol technologies. In: IBMA (Ed.) International Biocontrol Manufacturers Association. Available at: https://ibma-global.org/wp-content/uploads/2020/12/ibmadefinitionleafletweb.pdf.

Iñiguez-Moreno, M., Ragazzo-Sánchez, J. A., Barros-Castillo, J. C., Sandoval-Contreras, T. and Calderón-Santoyo, M. 2020. Sodium alginate coatings added with *Meyerozyma*

© Burleigh Dodds Science Publishing Limited, 2022. All rights reserved.

caribbica: postharvest biocontrol of *Colletotrichum gloeosporioides* in avocado (*Persea americana* Mill. cv. Hass). *Postharvest Biology and Technology* 163, 111123.

Kalantari, S., Marefat, A., Naseri, B. and Hemmati, R. 2018. Improvement of bean yield and Fusarium root rot biocontrol using mixtures of *Bacillus*, *Pseudomonas* and *Rhizobium*. *Tropical Plant Pathology* 43(6), 499–505.

Kamal, M. M., Lindbeck, K. D., Savocchia, S. and Ash, G. J. 2015. Biological control of Sclerotinia stem rot of canola using antagonistic bacteria. *Plant Pathology* 64(6), 1375–1384.

Kehe, J., Kulesa, A., Ortiz, A., Ackerman, C. M., Thakku, S. G., Sellers, D., Kuehn, S., Gore, J., Friedman, J. and Blainey, P. C. 2019. Massively parallel screening of synthetic microbial communities. *Proceedings of the National Academy of Sciences of the United States of America* 116(26), 12804–12809.

Koch, E., Becker, J. O., Berg, G., Hauschild, R., Jehle, J., Kohl, J. and Smalla, K. 2018. Biocontrol of plant diseases is not an unsafe technology! *Journal of Plant Diseases and Protection* 125(2), 121–125.

Köhl, J., Booij, K., Kolnaar, R. and Ravensberg, W. J. 2019a. Ecological arguments to reconsider data requirements regarding the environmental fate of microbial biocontrol agents in the registration procedure in the European Union. *BioControl* 64(5), 469–487.

Köhl, J., Kolnaar, R. and Ravensberg, W. J. 2019b. Mode of action of microbial biological control agents against plant diseases: relevance beyond efficacy. *Frontiers in Plant Science* 10, 845. doi: 10.3389/fpls.2019.00845.

Köhl, J., Medeiros, F. H. V., Lombaers-Van Der Plas, C., Groenenboom-De Haas, L. and Van Den Bosch, T. 2020. Efficacies of bacterial and fungal isolates in biocontrol of *Botrytis cinerea* and *Pseudomonas syringae* pv. *tomato* and growth promotion in tomato do not correlate. *Biological Control* 150, 104375.

Köhl, J., Postma, J., Nicot, P., Ruocco, M. and Blum, B. 2011. Stepwise screening of microorganisms for commercial use in biological control of plant-pathogenic fungi and bacteria. *Biological Control* 57(1), 1–12.

Kong, Q. 2018. Marine microorganisms as biocontrol agents against fungal phytopathogens and mycotoxins. *Biocontrol Science and Technology* 28(1), 77–93.

Korsten, L. 2004. Biological control in Africa: can it provide a sustainable solution for control of fruit diseases? *South African Journal of Botany* 70(1), 128–139.

Lamichhane, J. R., Bischoff-Schaefer, M., Bluemel, S., Dachbrodt-Saaydeh, S., Dreux, L., Jansen, J. P., Kiss, J., Köhl, J., Kudsk, P., Malausa, T., Messéan, A., Nicot, P. C., Ricci, P., Thibierge, J. and Villeneuve, F. 2017. Identifying obstacles and ranking common biological control research priorities for Europe to manage most economically important pests in arable, vegetable and perennial crops. *Pest Management Science* 73(1), 14–21.

Landa, B. B., Navas-Cortés, J. A. and Jiménez-Díaz, R. M. 2004. Influence of temperature on plant-rhizobacteria interactions related to biocontrol potential for suppression of fusarium wilt of chickpea. *Plant Pathology* 53(3), 341–352.

Lecompte, F., Abro, M. A. and Nicot, P. C. 2010. Contrasted responses of *Botrytis cinerea* isolates developing on tomato plants grown under different nitrogen nutrition regimes. *Plant Pathology* 59(5), 891–899.

Lecompte, F., Abro, M. A. and Nicot, P. C. 2013. Can plant sugars mediate the effect of nitrogen fertilization on lettuce susceptibility to two necrotrophic pathogens: *Botrytis cinerea* and *Sclerotinia sclerotiorum*? *Plant and Soil* 369(1–2), 387–401.

© Burleigh Dodds Science Publishing Limited, 2022. All rights reserved.

Leus, L. 2018. Breeding for disease resistance in ornamentals. In: Van Huylenbroeck J. (Ed.) Ornamental Crops. Handbook of Plant Breeding, Vol. 11, pp. 97–125. Cham: Springer. https://doi.org/10.1007/978-3-319-90698-0_5

Lombardi, N., Vitale, S., Turrà, D., Reverberi, M., Fanelli, C., Vinale, F., Marra, R., Ruocco, M., Pascale, A., D'errico, G., Woo, S. L. and Lorito, M. 2018. Root exudates of stressed plants stimulate and attract Trichoderma soil fungi. *Molecular Plant-Microbe Interactions: MPMI* 31(10), 982–994.

Lorito, M., Woo, S. L., Harman, G. E. and Monte, E. 2010. Translational research on Trichoderma: from'omics to the field. *Annual Review of Phytopathology* 48, 395–417.

Malerba, M. and Cerana, R. 2019. Recent applications of chitin- and chitosan-based polymers in plants. *Polymers* 11(5), 839.

Marian, M. and Shimizu, M. 2019. Improving performance of microbial biocontrol agents against plant diseases. *Journal of General Plant Pathology* 85(5), 329–336.

Marimuthu, S., Karthic, C., Mostafa, A. A., Mohammed Al-Enazi, N., Abdel-Raouf, N. and Nageh Sholkamy, E. 2020. Antifungal activity of *Streptomyces* sp. SLR03 against tea fungal plant pathogen *Pestalotiopsis theae*. *Journal of King Saud University–Science* 32(8), 3258–3264.

Marín, A., Atarés, L. and Chiralt, A. 2017. Improving function of biocontrol agents incorporated in antifungal fruit coatings: a review. *Biocontrol Science and Technology* 27(10), 1220–1241.

Marín, A., Plotto, A., Atarés, L. and Chiralt, A. 2019. Lactic acid bacteria incorporated into edible coatings to control fungal growth and maintain postharvest quality of grapes. *Hortscience* 54(2), 337–343.

Marrone, P. G. 2009. Barriers to adoption of biological control agents and biological pesticides. *CAB Reviews: Perspectives in Agriculture, Veterinary Science, Nutrition and Natural Resources* 2(51), 163–178.

Marrone, P. G. 2019. Pesticidal natural products–status and future potential. *Pest Management Science* 75(9), 2325–2340.

Massart, S., Martinez-Medina, M. and Jijakli, M. H. 2015. Biological control in the microbiome era: challenges and opportunities. *Biological Control* 89, 98–108.

Mauch-Mani, B., Baccelli, I., Luna, E. and Flors, V. 2017. Defense priming: an adaptive part of induced resistance. *Annual Review of Plant Biology* 68, 485–512.

Mazzola, M. and Freilich, S. 2017. Prospects for biological soilborne disease control: application of indigenous versus synthetic microbiomes. *Phytopathology* 107(3), 256–263.

Mitter, B., Brader, G., Pfaffenbichler, N. and Sessitsch, A. 2019. Next generation microbiome applications for crop production–limitations and the need of knowledge–based solutions. *Current Opinion in Microbiology* 49, 59–65.

Moser, R., Pertot, I., Elad, Y. and Raffaelli, R. 2008. Farmers' attitudes toward the use of biocontrol agents in IPM strawberry production in three countries. *Biological Control* 47(2), 125–132.

Mülner, P., Bergna, A., Wagner, P., Sarajlić, D., Gstöttenmayr, B., Dietel, K., Grosch, R., Cernava, T. and Berg, G. 2019. Microbiota associated with sclerotia of soilborne fungal pathogens–a novel source of biocontrol agents producing bioactive volatiles. *Phytobiomes Journal* 3(2), 125–136.

Nguyen, H. T. and Le, H. T. 2020. The effect of agricultural product eco-labelling on green purchase intention. *Management Science Letters* 10, 2813–2820.

© Burleigh Dodds Science Publishing Limited, 2022. All rights reserved.

Nicot, P., Alabouvette, C., Bardin, M., Blum, B., Kohl, J. and Ruocco, M. 2012. Review of factors influencing the success or failure of biocontrol: technical, industrial and socio-economic perspectives. *IOBC WPRS Bulletin* 78, 95–98.

Nicot, P. C., Bardin, M., Debruyne, F., Duffand, M., Lecompte, F., Neu, L. and Pascal, M. 2013. Effect of nitrogen fertilisation of strawberry plants on the efficacy of defence-stimulating biocontrol products against Botrytis cinerea. *IOBC WPRS Bulletin* 88, 39–42.

OECD 2009. OECD guidance document on defining minor uses of pesticides. ENV/JM/MONO(2009)39.

Ojiambo, P. S. and Scherm, H. 2006. Biological and application-oriented factors influencing plant disease suppression by biological control: a meta-analytical review. *Phytopathology* 96(11), 1168–1174.

Ozyilmaz, U. 2020. Evaluation of the effectiveness of antagonistic bacteria against Phytophthora blight disease in pepper with artificial intelligence. *Biological Control* 151, 104379.

Patel, J. S., Selvaraj, V., Gunupuru, L. R., Rathor, P. K. and Prithiviraj, B. 2020. Combined application of Ascophyllum nodosum extract and chitosan synergistically activates host-defense of peas against powdery mildew. *BMC Plant Biology* 20(1), 113.

Pelaez, V. and Mizukawa, G. 2017. Diversification strategies in the pesticide industry: from seeds to biopesticides. *Ciência Rural* 47(2). http://www.scielo.br/scielo.php?script=sci_arttext&pid=S0103-84782017000200152&nrm=iso.

Pintye, A., Ropars, J., Harvey, N., Shin, H. D., Leyronas, C., Nicot, P. C., Giraud, T. and Kiss, L. 2015. Host phenology and geography as drivers of differentiation in generalist fungal mycoparasites. *PLoS ONE* 10(3), e0120703.

Porat, R., Lichter, A., Terry, L. A., Harker, R. and Buzby, J. 2018. Postharvest losses of fruit and vegetables during retail and in consumers' homes: quantifications, causes, and means of prevention. *Postharvest Biology and Technology* 139, 135–149.

Postma, J. and Nijhuis, E. H. 2019. Pseudomonas chlororaphis and organic amendments controlling Pythium infection in tomato. *European Journal of Plant Pathology* 154(1), 91–107.

Poudel, R., Jumpponen, A., Schlatter, D. C., Paulitz, T. C., Gardener, B. B. M., Kinkel, L. L. and Garrett, K. A. 2016. Microbiome networks: a systems framework for identifying candidate microbial assemblages for disease management. *Phytopathology* 106(10), 1083–1096.

Rao, J., Chen, B. and McClements, D. J. 2019. Improving the efficacy of essential oils as antimicrobials in foods: mechanisms of action. *Annual Review of Food Science and Technology* 10, 365–387.

Raveau, R., Fontaine, J. and Sahraoui, A. L. H. 2020. Essential oils as potential alternative biocontrol products against plant pathogens and weeds: a review. *Foods* 9(3), 31.

Raymaekers, K., Ponet, L., Holtappels, D., Berckmans, B. and Cammue, B. P. A. 2020. Screening for novel biocontrol agents applicable in plant disease management–a review. *Biological Control* 144, 104240.

Robin, D. C. and Marchand, P. A. 2019. Evolution of the biocontrol active substances in the framework of the European pesticide Regulation (EC) No. 1107/2009. *Pest Management Science* 75(4), 950–958.

Roger, C., Tivoli, B. and Huber, L. 1999. Effects of temperature and moisture on disease and fruit body development of *Mycosphaerella pinodes* on pea (*Pisum sativum*). *Plant Pathology* 48(1), 1-9.

© Burleigh Dodds Science Publishing Limited, 2022. All rights reserved.

Romanazzi, G., Feliziani, E. and Sivakumar, D. 2019. Chitosan, a biopolymer with triple action on postharvest decay of fruit and vegetables: eliciting, antimicrobial and film-forming properties. *Frontiers in Microbiology* 9, 2745.

Rossi, V., Sperandio, G., Caffi, T., Simonetto, A. and Gilioli, G. 2019. Critical success factors for the adoption of decision tools in IPM. *Agronomy-Basel* 9, 19.

Sahu, P. K., Singh, S., Gupta, A. R., Gupta, A., Singh, U. B., Manzar, N., Bhowmik, A., Singh, H. V. and Saxena, A. K. 2020. Endophytic bacilli from medicinal-aromatic perennial Holy basil (*Ocimum tenuiflorum* L.) modulate plant growth promotion and induced systemic resistance against *Rhizoctonia solani* in rice (*Oryza sativa* L.). *Biological Control* 150, 104353.

SANCO 2017. Guidelines on comparability, extrapolation, group tolerances and data requirements for setting MRLs. *Guidance document*. European Commission, Directorate-General for Health and Food Safety.

Sarven, M. S., Hao, Q., Deng, J., Yang, F., Wang, G., Xiao, Y. and Xiao, X. 2020. Biological control of tomato gray mold caused by *Botrytis cinerea* with the entomopathogenic fungus *Metarhizium anisopliae*. *Pathogens* 9(3), 213.

Schlatter, D., Kinkel, L., Thomashow, L., Weller, D. and Paulitz, T. 2017. Disease suppressive soils: new insights from the soil microbiome. *Phytopathology* 107(11), 1284–1297.

Sharma, V., Salwan, R., Sharma, P. N. and Gulati, A. 2017. Integrated translatome and proteome: approach for accurate portraying of widespread multifunctional aspects of Trichoderma. *Frontiers in Microbiology* 8, 1602.

Shtienberg, D. and Elad, Y. 1997. Incorporation of weather forecasting in integrated, biological-chemical management of Botrytis cinerea. *Phytopathology* 87(3), 332–340.

Singh, S., Kumar, V., Dhanjal, D. S. and Singh, J. 2020. Biological control agents: diversity, ecological significances, and biotechnological applications. In: Singh, J. and Yadav, A. N. (Eds) *Natural Bioactive Products in Sustainable Agriculture*, pp. 31–44. Singapore: Springer.

Smith, K. P., Handelsman, J. and Goodman, R. M. 1999. Genetic basis in plants for interactions with disease-suppressive bacteria. *Proceedings of the National Academy of Sciences of the United States of America* 96(9), 4786–4790.

Sood, M., Kapoor, D., Kumar, V., Sheteiwy, M. S., Ramakrishnan, M., Landi, M., Araniti, F. and Sharma, A. 2020. Trichoderma: the "secrets" of a multitalented biocontrol agent. *Plants-Basel* 9(6), 25.

Tao, C., Li, R., Xiong, W., Shen, Z., Liu, S., Wang, B., Ruan, Y., Geisen, S., Shen, Q. and Kowalchuk, G. A. 2020. Bio-organic fertilizers stimulate indigenous soil Pseudomonas populations to enhance plant disease suppression. *Microbiome* 8(1), 137.

Thøgersen, J., Pedersen, S. and Aschemann-Witzel, J. 2019. The impact of organic certification and country of origin on consumer food choice in developed and emerging economies. *Food Quality and Preference* 72, 10–30.

Tucci, M., Ruocco, M., De Masi, L., De Palma, M. and Lorito, M. 2011. The beneficial effect of Trichoderma spp. on tomato is modulated by the plant genotype. *Molecular Plant Pathology* 12(4), 341–354.

Van Eeden, M. and Korsten, L. 2003. Determining appropriate parameters for an integrated disease control system for biological predictive modelling. *South African Avocado Growers' Association Yearbook* 26, 83–95.

© Burleigh Dodds Science Publishing Limited, 2022. All rights reserved.

van Lenteren, J. C., Bolckmans, K., Köhl, J., Ravensberg, W. J. and Urbaneja, A. 2018. Biological control using invertebrates and microorganisms: plenty of new opportunities. *BioControl* 63(1), 39–59.

van Lenteren, J. C. and Nicot, P. C. 2020. Integrated pest management methods and considerations concerning implementation in greenhouses. In: Gullino, M. L., Albajes, R. and Nicot, P. C. (Eds) *Integrated Pest and Disease Management in Greenhouse Crops*, pp. 1–20. Cham: Springer International Publishing. Available at: https://www.sciencedirect.com/science/article/abs/pii/S1369703X07001994.

Verma, M., Brar, S. K., Tyagi, R. D., Surampalli, R. Y. and Valero, J. R. 2007. Antagonistic fungi, Trichoderma spp.: panoply of biological control. *Biochemical Engineering Journal* 37(1), 1–20.

Villemaine, R., Compagnone, C. and Falconnet, C. 2020. The social construction of alternatives to pesticide use: A study of biocontrol in Burgundian viticulture. *Sociologia Ruralis* 61(1), 74–95, doi: 10.1111/soru.12320.

Vitorino, L. C. and Bessa, L. A. 2018. Microbial Diversity: the gap between the estimated and the known. *Diversity* 10(2), 29.

Wamalwa, L. N., Cheseto, X., Ouna, E., Kaplan, F., Maniania, N. K., Machuka, J., Torto, B. and Ghislain, M. 2015. Toxic ipomeamarone accumulation in healthy parts of sweetpotato (*Ipomoea batatas* L. Lam) storage roots upon infection by *Rhizopus stolonifer*. *Journal of Agricultural and Food Chemistry* 63(1), 335–342.

Wei, Z., Huang, J., Yang, T., Jousset, A., Xu, Y., Shen, Q. and Friman, V. P. 2017. Seasonal variation in the biocontrol efficiency of bacterial wilt is driven by temperature-mediated changes in bacterial competitive interactions. *Journal of Applied Ecology* 54(5), 1440–1448.

Woo, S. L. and Pepe, O. 2018. Microbial consortia: promising probiotics as plant biostimulants for sustainable agriculture. *Frontiers in Plant Science* 9, 1801.

Yuan, B., Jia, H., Bu, W., Yang, T., Han-Meng, L., Ju, X. Y. and Li, R. P. 2020. A new chitosan sub-micron and encapsulated Iturin A with enhanced antifungal activity against *Ceratocystis fimbriata* and *Rhizopus strolonifer*. *International Journal of Biological Macromolecules* 159, 995–1003.

Zaki, O., Weekers, F., Thonart, P., Tesch, E., Kuenemann, P. and Jacques, P. 2020. Limiting factors of mycopesticide development. *Biological Control* 144, 104220.

Zhai, Z., Martínez, J. F., Beltran, V. and Martínez, N. L. 2020. Decision support systems for agriculture 4.0: survey and challenges. *Computers and Electronics in Agriculture* 170, 105256.

Zhou, Y., Zhang, L. and Zeng, K. 2016. Efficacy of pichia membranaefaciens combined with chitosan against Colletotrichum gloeosporioides in citrus fruits and possible modes of action. *Biological Control* 96, 39–47.

© Burleigh Dodds Science Publishing Limited, 2022. All rights reserved.

Chapter 3

Effective biocontrol of grape diseases: assessing and improving efficacy

V. Altieri, P. Battilani, M. Camardo Leggieri, G. Fedele, T. Ji, V. Rossi and I. Salotti, Università Cattolica del Sacro Cuore, Italy

1 Introduction

Grapevine is one of the most important crops globally, accounting for more than 7.3 million hectares of cultivated area (Roca, 2022). Grapevine is affected

http://dx.doi.org/10.19103/AS.2024.0142.18
© Burleigh Dodds Science Publishing Limited, 2025. All rights reserved.

by several diseases which vary by region (Bois et al., 2017), with mildews being the major phytosanitary threat in most areas. These diseases negatively impact both grapevine yield and wine quality (Gouadec et al., 2007), resulting in significant economic losses (Van Helden, 2008). To maintain production and quality standards, intensive fungicide use is often required (Compant et al., 2013). For instance, although the area given to grapes accounts for just 3.5% of the total agricultural area in the European Union (EU), grapes receive around 15% of the synthetic pesticides (as active substances, a.s.) applied to major crops, with an average of 4.7 kg of a.s./ha between 1992 and 2003 (EUROSTAT, 2007). The extensive use of synthetic pesticides and awareness of their negative impact on human health (Loevinsohn, 1987) and the environment (Conway and Pretty, 1991) have led to efforts to limit their use. The use of synthetic fungicides pre-harvest has undergone limitations in order to reduce or eliminate pesticide residues on both grapes and wine (Verger and Boobis, 2013). The EU's Farm to Fork strategy aims for a 50% reduction in the overall use of chemicals and more hazardous active substances by 2030 (European Commission, 2020).

To minimize the negative impacts of pesticides, low-risk, alternative plant protection products are being considered and implemented in Integrated Pest Management and organic farming (Rossi et al., 2012; Pertot et al., 2017). These alternatives include biological control agents (BCAs), which are microorganisms (bacteria, oomycetes, and fungi) that are able to interact, directly or indirectly, with pathogens and to reduce disease development (Kumar and Droby, 2021; Elad and Freeman, 2002; Elmer and Reglinski, 2006). BCAs have minimal impact on human health and the environment, and have limited side effects on non-target organisms (Ritika and Utpal 2014; Parnell et al. 2016; Syed Ab Rahman et al. 2018). They also have short pre-harvest intervals (Marrone, 2007) and can be used in anti-resistant strategies due to their modes of action (MoA) compared to synthetic fungicides (Corkley et al., 2022). Although highly complex and diverse, and often coexisting in the same microorganism, modes of action of BCAs include antibiosis due to the production of enzymes and/or antibiotic metabolites, competition for space and nutrients, hyperparasitism of the pathogen, and induction of resistance in plants (Pertot et al., 2016). However, BCAs often have lower and more variable efficacy compared to synthetic fungicides, especially when applied under field conditions (Tracy, 2014; Pertot et al., 2017; Altieri et al., 2023).

Although several studies have been carried out to elucidate the mode of action of the different BCAs (e.g. Compant et al., 2013; Velivelli et al., 2014; Raymaekers et al., 2020, Hamdi and Alzawi, 2023), research on their effectiveness remains fragmented, making it difficult to draw clear and robust conclusions.

© Burleigh Dodds Science Publishing Limited, 2025. All rights reserved.

2 Current research on biocontrol of grape diseases: an overview

To estimate the amount of information available on biocontrol of grape disease, a Google search was carried out on 24 July 2023, using the string "biocontrol disease grape", to ensure that all the terms appeared. The search returned over 1.5 million results. When the search was repeated on Google Scholar (using the same string and searching for any type of publication), it yielded 31,200 results. When the search was restricted to scientific papers, this number decreased to approximately 6,400 results. The search results were then segmented by year (Figure 1), revealing that fewer than 50 papers were published in every five-year period until 1985, whereafter the number increased exponentially to more than 2,000 published between 2021 and July 2023, including 523 papers published in the first seven months of 2023. These figures clearly demonstrate a growing interest in the study of BCAs for the control of grape diseases, beginning in the late 1980s.

When the search for scientific papers was further segmented to focus on grape diseases with worldwide distribution (Bois et al., 2017) (Figure 2), the following results were found:

- 55% (2,340 papers) mentioned grey mould, caused by *Botrytis cinerea* (using the string "biocontrol botrytis grape").
- 15.3% (653 papers) mentioned trunk diseases (which include different diseases and causal agents, as outlined in Mondello et al., 2018).

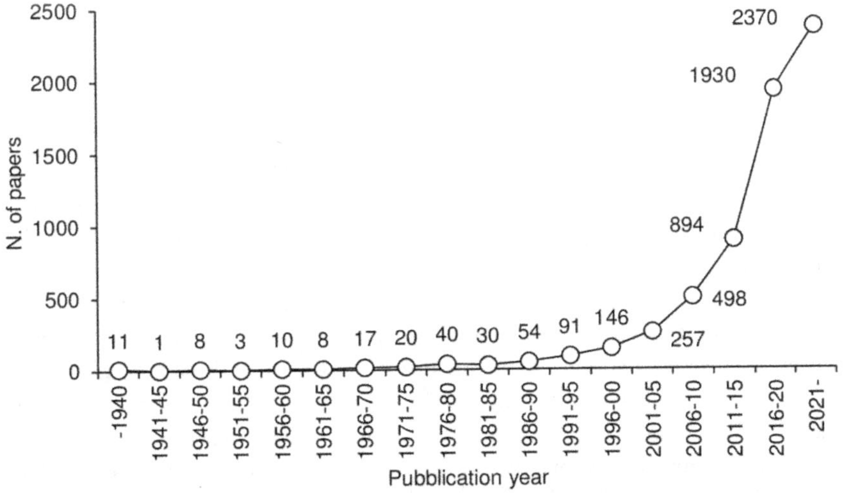

Figure 1 Number of papers published on the biocontrol of grape diseases over time, as obtained through a search on Google Scholar. Excludes references to papers published before 1940 and extrapolates papers for every five-year period until July 2023.

© Burleigh Dodds Science Publishing Limited, 2025. All rights reserved.

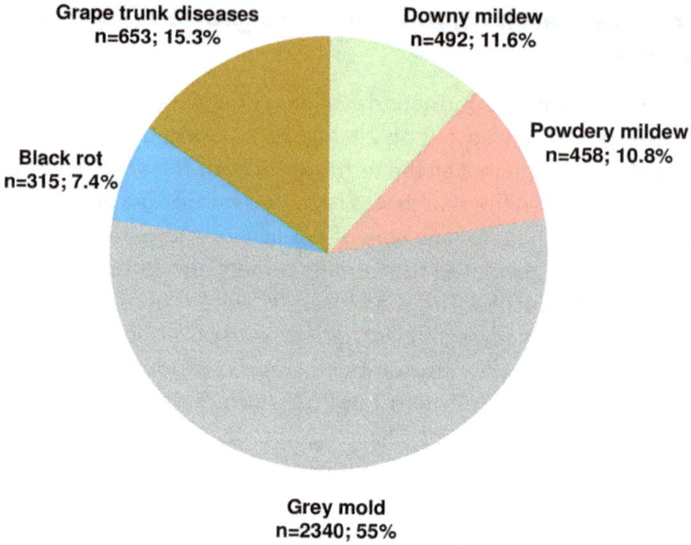

Figure 2 Number of papers published on the biocontrol of major grape diseases up to July 2023, as obtained through a search on Google Scholar.

- 11.6% (492 papers) and 10.8% (458 papers) mentioned downy mildew (caused by *Plasmopara viticola*) and powdery mildew (caused by *Erysiphe necator*), respectively.
- 7.4% (653 papers) mentioned black rot (caused by *Phyllosticta ampelicida*).

Papers mentioning white rot (caused by *Coniella diplodiella*) were few (numbering just 18), unlike those mentioning Aspergillus rot (caused by *Aspergillus* spp.) and anthracnosis (caused by *Elsinoë ampelina*). Surprisingly, 2,270 papers mentioned bunch rots caused by *Aspergillus* (using the search term "biocontrol aspergillus grape"), and 726 papers included both the terms "aspergillus" and "ochratoxin", i.e. the mycotoxin Ochratoxin A (OTA) produced by black *Aspergilla* and particularly *A. carbonarius* (Perrone et al., 2004; Battilani et al., 2006). This result is somewhat skewed by the fact that the term "grape" or the disease of interest was mentioned only in the introduction to the paper, with the actual research focusing on other crops, food products or pathogens. A similar situation was observed for *Colletorichum* rot ("biocontrol colletortichum grape") and sour rot (caused by multiple microorganisms), which were mentioned in 1,670 and 458 papers, respectively.

With the exception of rots caused by *Aspergillus* and *Colletorichum* species (as discussed above), there is consistency between the global relevance of grape diseases and the number of papers on biocontrol, with the exception of grey mould (GM). The extensive research into GM may be

© Burleigh Dodds Science Publishing Limited, 2025. All rights reserved.

attributed to the fact that avoidance of the disease necessitates repeated fungicide applications during berry ripening. There is an urgent requirement for alternative solutions to synthetic fungicides to counter the problems related to the pre-harvest interval (i.e. the minimum number of days between the last pesticide application and the harvest) and to pesticide residues in wine (i.e. the pesticide that may remain on or in berries after they are applied to the crop). *Aspergillus* and *Colletorichum* bunch rots suffer from the same issues as GM (De Curtis et al., 2012; Pedrotti et al., 2020). This is not the case for mildews and black rot, where berries develop so-called ontogenetic resistance (age-related resistance; Gadoury et al., 2003, Kennely et al., 2005; Molitor Berkelmann-Loehnertz, 2011), and no specific pesticide applications are required to protect grape clusters from pre-veraison onwards (Bleyer et al., 2023). Intensive fungicide control is less critical for these diseases compared to GM, and effective alternatives to synthetic pesticides exist, including copper and sulphur (Rantsiou et al., 2020). Growing environmental concerns over the side effects of these inorganic compounds (Conway and Pretty, 1991; Hinckley et al., 2020) and subsequent restrictions in their usage in some countries (e.g. Directive 2009/128 in the European Community), have fueled interest in alternative solutions, including BCAs (Pertot et al., 2017). Interest in biocontrol for trunk diseases stems from the lack (or scarcity) of fungicides able to efficiently prevent plants from being infected by causal agents of wood decay and/or vascular colonization (Mondello et al., 2018). These pathogens typically enter the plant through wounds, especially those made during pruning (Gramaje et al., 2018).

The literature identified via the Google search covers a wide range of topics, including:

- Identification and selection of putative biocontrol microorganisms (e.g. Cordero-Bueso et al., 2017; Bruisson et al., 2019).
- Screening for efficacy against target pathogens in the laboratory or under controlled conditions (e.g. Niem et al. 2020; Zhao et al., 2022).
- Studies into their MoA, such as hyperparasitism (e.g. Angeli et al., 2009, 2012), production of toxic metabolites (e.g. Leifert et al., 1995; Castoria et al., 2001; Toral et al., 2018), competition (e.g. Sipiczki, 2006) and/or induced resistance in plants (e.g. Mohamed et al., 2007; Choudhary and Johri, 2009), using both classical (e.g. Nally et al., 2015) and molecular (e.g. Niem et al, 2023) methods.
- Evaluation of their effectiveness in disease control, either alone or in combination with BCAs and fungicides, under controlled conditions (e.g. Raut et al., 2018; Zhou et al., 2020) and in the vineyard (e.g. Calvo-Garrido et al., 2013; Sawant et al., 2017), using both artificial (e.g. Brown et al., 2021) and natural (e.g. Falk et al., 1996) pathogen inoculum.

© Burleigh Dodds Science Publishing Limited, 2025. All rights reserved.

This chapter focuses on the efficacy of BCAs, with the goal of critically and quantitatively examining the effectiveness of biocontrol for major grapevine diseases across the following studies:

- Downy mildew (DM);
- Powdery mildew (PM);
- Grey mould (GM);
- Black rot (BR);
- Trunk diseases (TDs; specifically Botryosphaeria dieback, Esca complex and Petri disease, Eutypa dieback, and black foot, as described in Mondello et al., 2018);
- Bunch rots caused by ochratoxigenic fungi (OF), i.e. fungi producing Ochratoxin A (OTA); and
- Sour rot (SR).

The aim is to (i) draw conclusions on the use of BCAs and (ii) identify knowledge gaps to be addressed by future research for more sustained use of biocontrol in vineyards.

3 Assessing the efficacy of biological control agents (BCAs) against grapevine diseases

To review the available literature on the efficacy of biocontrol, a second search was conducted using international bibliographical databases and a systematic literature approach. Systematic literature reviews are an essential aspect of academic research as they build on existing evidence to advance knowledge (Xiao and Watson, 2019). A systematic literature review provides a thorough evaluation of a research topic through a reliable, rigorous and verifiable methodology that involves summarizing, analyzing and synthesizing the related literature (Keele, 2007). A systematic approach reduces errors, limits researcher bias and communication of the information is enhanced (Candel, 2014).

3.1 Systematic literature review

The literature search was conducted across three databases:

- Scopus (https://www.scopus.com; accessed on 1 February 2023).
- Web of Science (https://www.webofscience.com; accessed on 1 February 2023).
- China National Knowledge Infrastructure (https://www.cnki.net/index/; accessed on 2 February 2023).

© Burleigh Dodds Science Publishing Limited, 2025. All rights reserved.

In order to be included in the database of relevant literature, papers had to satisfy the following pre-defined inclusion and exclusion criteria (Okoli and Schabram, 2010; Candel, 2014):

- The microorganism had to be identified at the taxonomic level and used as a strain or a commercial product.
- The microorganism had to be considered a BCA (i.e. studies focusing on the biology or ecology of the microorganism, or its MoA, were excluded).
- The BCA had to be tested for its ability to control at least one of the targeted diseases/pathogens on grape plants or organs, under controlled or field conditions. Studies conducted in the lab in which the microorganism and/or the pathogen were grown on artificial media were excluded, with the exception of OF.
- The scheduling strategy considered the application of a single BCA only (i.e. strategies that combined different BCAs and/or chemical products were not considered).
- The results needed to be expressed as efficacy relative to untreated control, or the study's data (e.g. disease incidence or severity, OTA concentration) had to allow for the calculation of efficacy.

Based on these criteria, specific queries were formulated by combining the following keywords:

- Biocontrol agent OR biological control agent OR biological control OR biocontrol OR BCA.
- Grape (and its multiple endings) OR vineyard OR *Vitis*.
- The scientific name of the pathogen OR synonyms OR the name of the disease OR other common names, e.g. botrytis bunch rot OR grey mould OR gray mold OR Botrytis bunch rot OR *Botrytis cinerea* OR *Botryotinia fuckeliana* (see disease-specific keywords in Table 1).

The data collection process is schematically described in Figure 3 (Biesbroek et al., 2013; Koricheva and Gurevitch, 2014). Papers obtained from the first search were merged and duplicates excluded; titles and abstracts were then screened to select the papers fulfilling the aim of this chapter, i.e. the *in planta* efficacy of BCAs used for the control of major grapevine diseases. Full texts of papers considered of potential interest were reviewed to ensure relevance. The numbers of screened and selected papers for each disease are reported in Table 1.

The systematic literature review yielded 176 papers in total, conducted between 1988 and 2022, in 32 countries that include Europe, namely Italy (32 studies), Spain (19), France (11) and Greece (11); America, specifically the US

© Burleigh Dodds Science Publishing Limited, 2025. All rights reserved.

Table 1 Keywords related to the investigated diseases and the corresponding number of papers found and selected for quantitative analysis.

Disease	Keywords	N° hits	N° selected for systematic
Downy mildew	Downy mildew OR *Plasmopara viticola*	212	26
Powdery mildew	Powdery mildew OR White mould OR *Erysiphe necator*	142	13
Black Rot	Black rot OR *Guignardia bidwellii* OR *Phyllosticta ampelicida*	15	0
Grapevine Trunk Diseases	Grape trunk diseases OR GTD OR Esca OR *Phaeoacremonium* OR *Phaeomoniella* OR *Eutypa* OR *Cadophora* OR *Botryospheria* OR *Diplodia* OR *Fomitiporia*	200	37
Grey Mould	Botrytis bunch rot OR grey mould OR gray mould OR grey mold OR gray mold OR *Botrytis cinerea* OR *Botryotinia fuckeliana*	622	83
Sour Rot	Sour rot	28	1
Ochratoxigenic Fungi	*Aspergillus* OR Black mould OR OTA OR OTB OR OTC OR ochratoxin	206	25

(11), Chile (5), Brazil (3) and Canada (2); China (26), India (10) and other Asian or Middle Eastern countries (5); South Africa (7) and Tunisia (5); Australia (3) and New Zealand (2). The selected papers referenced DM (26 papers), PM (13), GM (83), TDs (27), OF (17) and SR (1). No studies on the *in planta* biocontrol efficacy against BR were found.

3.2 Extraction of efficacy data from the literature

Data on BCA efficacy or disease assessments useful for calculating efficacy were extracted from the text, tables, and figures of the collected papers. The GetData Graph Digitizer 2.24 (http://getdata-graph-digitizer.com) was used to obtain precise data from graphs. Measured variables (X; i.e. disease incidence and severity, and OTA concentration) were used to calculate biocontrol efficacy (E) as follows: $E = (X_{NT} - X_T) / X_{NT}$, where NT represents the value of the untreated control and T is the value in any specific BCA-pathogen interaction (e.g. PM disease severity in a specific vineyard and year treated with *Bacillus subtilis*).

The extracted data were organized into a database and used to conduct a quantitative analysis of pooled data, enhancing their relevance compared to individual studies and aiding in summarizing the available scientific knowledge (Noordzij et al., 2009). This method is increasingly being used in plant pathology (Madden and Paul, 2011; Scherm et al., 2014; Ji et al., 2023; Rosace et al., 2023).

© Burleigh Dodds Science Publishing Limited, 2025. All rights reserved.

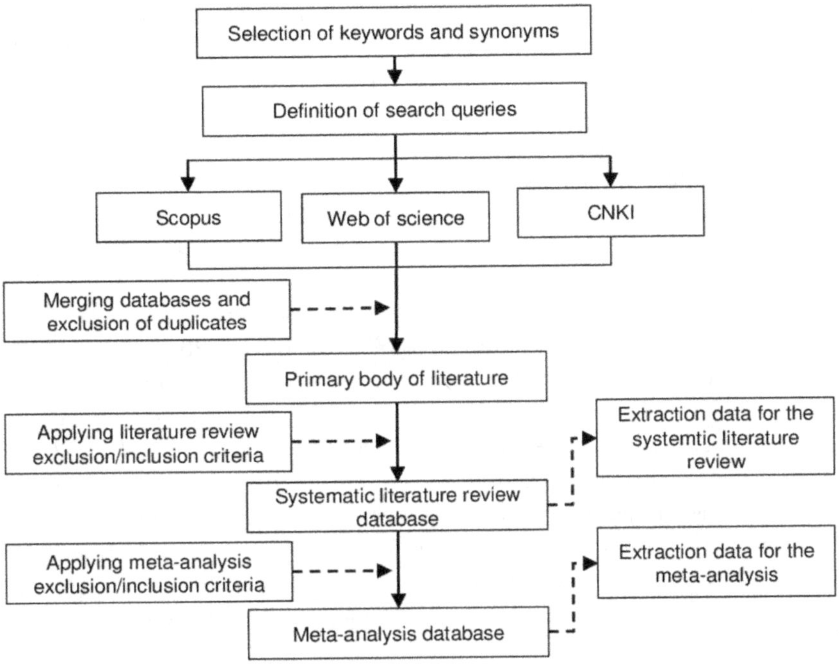

Figure 3 Schematic representation of the systematic literature review and data extraction process. Three international databases were used for the search: Scopus, Web of Science, China National Knowledge Infrastructure (CNKI).

It is necessary to note that the selected papers met the specific criteria outlined earlier (e.g. microorganisms were used as BCAs, efficacy data were available or calculable, BCAs were applied alone and not in combination or alternation with other plant protection products), and were published in the referenced databases. The analysis outlined in the following sections does not include all the papers published to date on the topic.

4 Biocontrol efficacy against downy mildew

A total of 26 papers on biocontrol efficacy on leaves and/or clusters were identified, which provided 138 efficacy figures. In half of these studies (13), BCAs were applied under controlled-environment conditions only, six papers focused on field conditions only, and seven papers included both controlled and field settings. The evaluation of BCA efficacy against DM was largely carried out on leaves, with only a few field studies assessing the effects on clusters.

Under controlled conditions, BCAs were typically applied on potted grapevine plants one to seven days before artificial inoculation with the pathogen. Detached (in laboratory) or attached (in greenhouse) leaves were mostly spray-inoculated with a *P. viticola* sporangia suspension (typically 10^5

© Burleigh Dodds Science Publishing Limited, 2025. All rights reserved.

sporangia/mL) and incubated at room temperature (20 to 25°C) and high relative humidity (95 to 100% RH) to ensure infection. Following the onset of DM, disease severity was assessed on leaves and/or clusters by using different diagrammatic scales (e.g. Banani et al., 2014; Jaber, 2015). Under field conditions, 9 of the 13 studies were conducted in the presence of natural inoculum; in these conditions, BCAs were applied twice or repeatedly during the season, usually on a calendar basis. Biocontrol efficacy was then assessed seven to 10 days after each application or after the final one (e.g. Falk et al., 1996). In studies where *P. viticola* was artificially inoculated in the field, BCA application, inoculum characteristics, and disease assessment were similar to that used under controlled conditions (e.g. Dagostin et al., 2011).

A total of 24 microorganisms were considered, belonging to 11 genera or singleton species of bacteria, filamentous fungi, and yeasts (Figure 4). Most BCAs were ascomycetous fungi (59% of papers), with the genera *Trichoderma* (39%; mainly *T. harzianum*) and *Fusarium* (35%; mainly *F. proliferatum*) represented more than the singleton species *Beauveria bassiana* (9%), *Saccharomyces cerevisiae* (9%), *Aureobasidium pullulans* (4%), and *Leptosphaerulina australis* (4%). Bacteria accounted for 41% of the papers, with *Bacillus* the most studied genus (56%, mainly *B. subtilis*) followed by *Streptomyces* (16%)

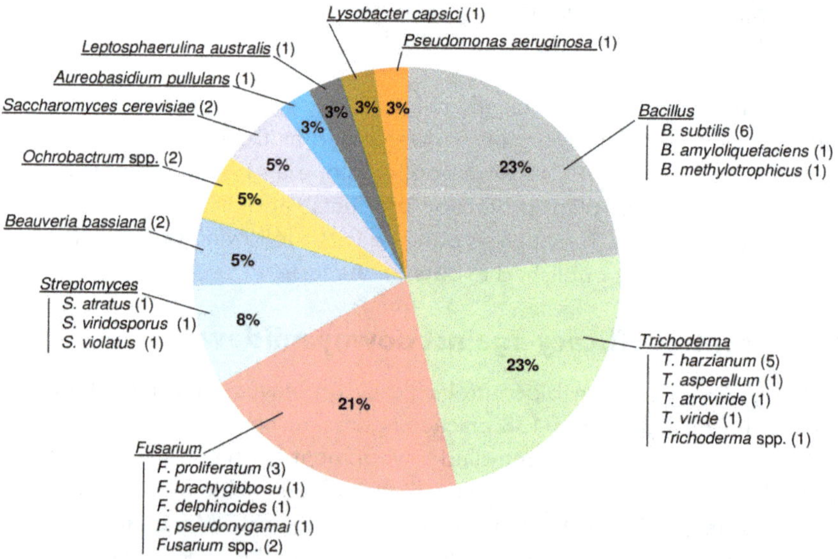

Figure 4 Representation of the microorganisms (singleton species or genera) for which papers on the *in planta* biocontrol efficacy against grape downy mildew have been found through a systematic literature review. Sectors show the amount of papers as a percentage of total papers, while the number of papers for each microorganism is reported in brackets.

© Burleigh Dodds Science Publishing Limited, 2025. All rights reserved.

and *Ochrobactrum* (13%). The singleton species *Lysobacter capsici* and *Pseudomonas aeruginosa* were studied in only one case.

Figure 5 shows the calculated efficacy of BCAs using disease severity data relative to untreated control, showing significant variability depending on the BCA genera and the test conditions. Under controlled conditions, the highest efficacy was observed for *Streptomyces* (average 92.1%; *n* = 1), *P. aeruginosa* (91.3%; *n* = 1), and *B. bassiana* (80.4 ± 2.8%; *n* = 8). Unfortunately, no field data were available for *P. aeruginosa*, while *Streptomyces* spp. and *B. bassiana* were tested only on leaves in the field. A reduction in efficacy was observed when transitioning from controlled to field conditions: from 92.1% to 68.8% (± 4.2%; *n* = 5) for *Streptomyces*, and from 80.4% to 51.0% (± 8.8%; *n* = 6) for *B. bassiana*. *Bacillus* showed high efficacy on leaves under controlled conditions (69.6 ± 9.5%; *n* = 13) and on clusters in the field (77.0 ± 12.5%; *n* = 3) but performed poorly on leaves in the field (44.5 ± 2.1%; *n* = 17). In contrast, *Trichoderma* exhibited higher efficacy under controlled conditions (59.6 ± 4.5%; *n* = 21) and on leaves in the field (52.4 ± 10.1%; *n* = 5), with negligible biocontrol on clusters in the field (23%; *n* = 1). The efficacy showed by *Fusarium* was slightly higher in the field (59.8 ± 7.8%; *n* = 11, and 66.6 ± 2.3%; *n* = 9, on

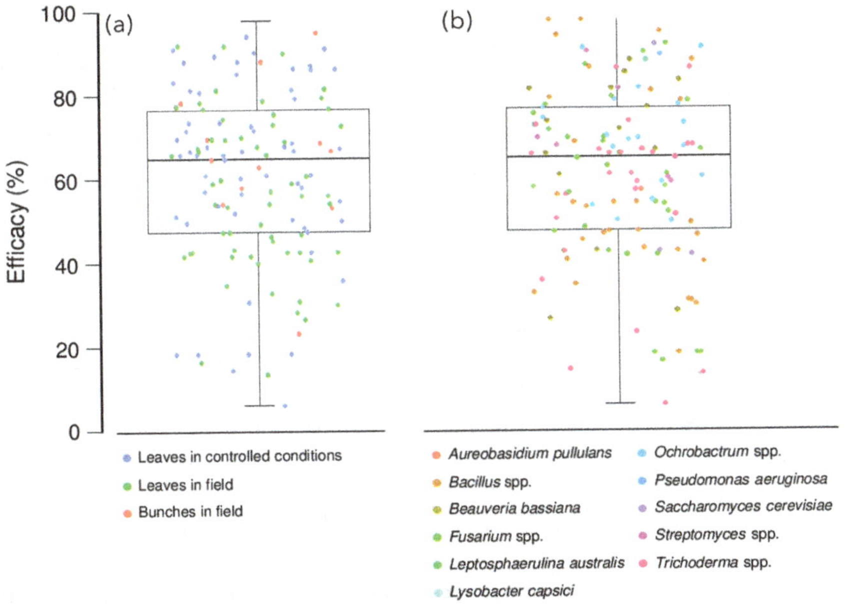

Figure 5 Box plots showing efficacy (%) data on *in planta* downy mildew biocontrol extracted from the papers shown in Figure 4. In panel A, data are grouped based on experimental conditions; in panel B they are based on microorganisms. Efficacy was calculated using disease severity data in relation to an untreated control. The box extends from the 25th to 75th quartile of the data distribution, and the line crossing the box represents the median; whiskers extend to the maximum and minimum.

© Burleigh Dodds Science Publishing Limited, 2025. All rights reserved.

leaves and clusters, respectively) than in controlled conditions (49.9 ± 7.0%; *n* = 11). The highest efficacy in field was observed for *L. capsici* by Markellou et al. (2022), with 74.0% (*n* = 1) and 90.0% (*n* = 1) on leaves and clusters, respectively; additional studies, however, are needed to confirm the promising performance of this microorganism. The efficacy of *A. pullulans* (*n* = 1), *L. australis* (*n* = 1), *Ochrobactrum* (*n* = 19), and *S. cerevisiae* (*n* = 3) ranged from 58% to 72.9% on leaves in both controlled and field conditions.

5 Biocontrol efficacy against powdery mildew

A review of the literature identified 13 papers on biocontrol efficacy against *E. necator* on leaves and/or clusters, resulting in 203 efficacy figures. Of these studies, nine focused exclusively on field conditions, three on controlled-environment conditions only, and one paper included both controlled and field conditions. The evaluation of BCA efficacy under controlled conditions was carried out exclusively on leaves, while the field studies assessed both the leaves and clusters.

Under controlled conditions, the control provided by BCAs was assessed *in planta*, typically grown in greenhouses. BCAs were applied at least twice, either before or after artificial inoculations were carried out by dusting *E. necator* conidia (Maachia et al., 2015; Konstantinidou-Doltsinis et al., 2007) or via natural inoculum present on greenhouse plants infected by PM (Crisp et al., 2006; Pascale et al., 2017). Disease severity was assessed on leaves at 10 to 20 days after inoculation. In field studies, BCAs were applied at regular intervals to control the disease originating from the natural inoculum. The provided control was assessed in terms of disease severity during the season and/or after the last BCA application (e.g. Ahmed, 2018; Falk et al., 1995; Konstantinidou-Doltsinis et al., 2007).

A total of 22 microorganisms were evaluated, representing five genera or individual species of bacteria and filamentous fungi (Figure 6). Most (59%) of the tested BCAs were ascomycetous fungi, with the genus *Trichoderma* (45%; mainly *T. harzianum*) more frequently considered than the singleton species *Ampelomyces quisqualis* (9%) and *Pseudozyma flocculosa* (5%). Bacteria accounted for 41% of papers, with *Bacillus subtilis* being the most common species (27%), followed by *Pseudomonas fluorescens* and *P. syringae* (14% combined).

Figure 7 illustrates the boxplot of BCAs' efficacy calculated via disease severity data with reference to an untreated control. In panel A, efficacy data are grouped based on the experiment (in lab or field) and the organ (leaves or clusters). In panel B, they are grouped based on the BCA genus. The boxplots and data dispersion reveal significant variability in the biocontrol efficacy data extracted from the different studies, although some trends can be identified.

© Burleigh Dodds Science Publishing Limited, 2025. All rights reserved.

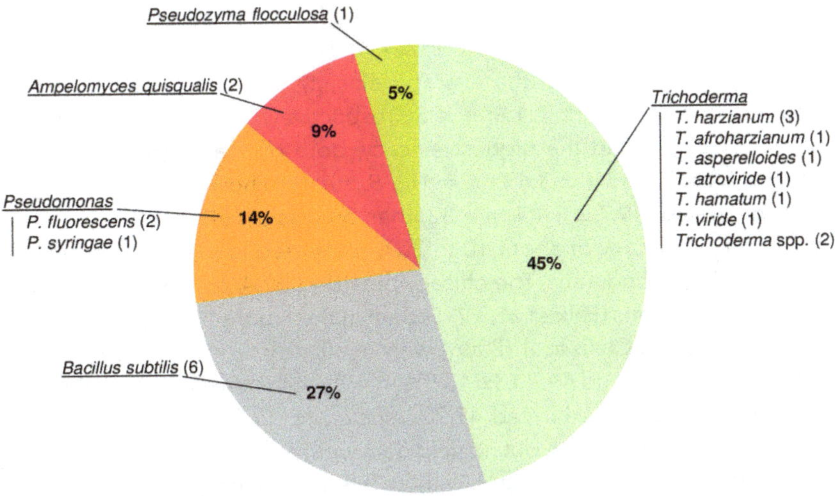

Figure 6 Representation of the microorganisms (singleton species or genera) for which papers on the *in planta* biocontrol efficacy against grape powdery mildew have been found through a systematic literature review. Sectors show the amount of papers as a percentage of total papers, while the number of papers for each microorganism is reported in brackets.

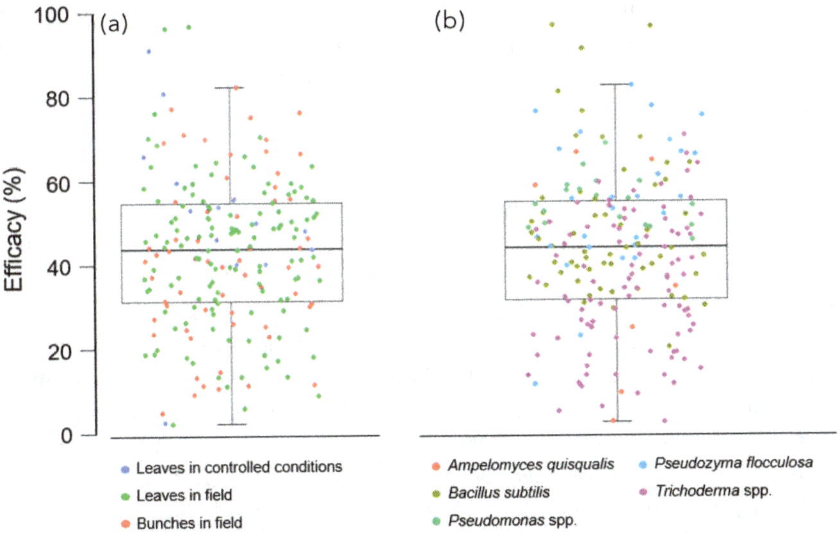

Figure 7 Box plots showing the efficacy (%) data on *in planta* powdery mildew biocontrol extracted from the papers shown in Figure 6. In panel A, data are grouped based on experimental conditions; in panel B they are based on microorganisms. Efficacy was calculated using disease severity data in relation to an untreated control. The box extends from the 25th to 75th quartile of the data distribution, and the line crossing the box represents the median; whiskers extend to the maximum and minimum.

© Burleigh Dodds Science Publishing Limited, 2025. All rights reserved.

Under controlled conditions, the highest efficacy on leaves was observed for *B. subtilis* (average of 60.8 ± 9.0%; *n* = 6). For *P. flocculosa, Pseudomonas* spp. and *Trichoderma* spp., the average efficacy was 55.2 (± 6.6%; *n* = 3), 54.0% (*n* = 1) and 49.1 (± 2.1%; *n* = 3), respectively. Under field conditions, *Pseudomonas* spp. had the highest efficacy, both on leaves (54.9 ± 1.5%; *n* = 18) and clusters (55.9 ± 3.4%; *n* = 18). *B. subtilis* showed a slightly higher efficacy on leaves (49.7 ± 3.1%; *n* = 32) than on clusters (42.6 ± 2.8%; *n* = 20), with a reduced efficacy of about 10 to 20% compared to the control obtained under controlled conditions. The efficacy provided by *A. quisqualis* was higher under field conditions (Falk et al., 1995) than in the laboratory experiment (1%; *n* = 1) conducted by Crisp et al. (2006), with an efficacy on leaves and clusters of 41% (± 12.3%; *n* = 3) and 44.7% (± 18.5%; *n* = 3), respectively. In a single study, *Pseudozima flocculosa* exhibited 49.7% efficacy (± 11.2%; *n* = 6) on clusters under controlled conditions, but no field data were available for leaves. Overall, *Trichoderma* showed poor control, with an average efficacy of 36.3 (±1.9%; *n* = 73) and 20.3 (±2.3%; *n* = 16) on leaves and clusters in the field, respectively.

6 Biocontrol efficacy against grey mould

Grey mould is the most studied grape disease for the use of BCAs, with 83 studies on biocontrol efficacy on leaves and/or clusters, contributing 426 efficacy figures. BCAs were applied under controlled-environment conditions in 42 of these papers, in the field in 20 papers, and under both conditions in 17 papers. Evaluation of BCA efficacy against GM was largely carried out on clusters.

Under controlled conditions, studies were conducted on leaves (e.g. Trotel-Aziz et al., 2008) and wounded (e.g. Lemos et al., 2016) and non-wounded berries (e.g. Fedele et al., 2020a) by applying BCAs as preventative (e.g. Maachia et al., 2015) or curative treatments (e.g. Whiteman and Stewart, 1998), a few hours or days before or after the artificial inoculation of a *B. cinerea* conidial suspension. Under field conditions, applications of BCAs were tested on clusters infected by natural (e.g. Calvo-Garrido et al., 2019) or artificial (e.g. Jelenić et al., 2021) inoculum. During these trials, BCAs were applied repeatedly at different timings. GM assessment was carried out by measuring disease incidence and/or severity in both environment-controlled and field conditions, via various assessment diagrams. In the field, assessments were typically carried out prior to harvesting (e.g. Magnin-Robert et al., 2013) and, in a few cases, after incubating field-collected berries with no disease symptoms in moist chambers for a week, to assess latent *B. cinerea* infections (e.g. Fedele et al., 2020b).

A total of 56 microorganisms were studied, belonging to 28 genera or singleton species of bacteria, filamentous fungi, and yeasts. An oomycete

© Burleigh Dodds Science Publishing Limited, 2025. All rights reserved.

was also considered, specifically *Pythium oligandrum* (Figure 8). Most of the tested BCAs were bacteria (51% of studies), with the genera *Bacillus* (27%; mainly *B. subtilis* and *B. amyloliquefaciens*) better represented than *Pantoea* (6%, especially *P. agglomerans*), *Pseudomonas* (5%, mainly *P. fluorescens*), *Enterobacter* (5%), *Paenibacillus* (3%), and singleton species accounting for under 3% each (i.e. *Acinetobacter lwoffii*, *Brevibacillus reuszeri*, *Burkholderia contaminans*, *Lactobacillus plantarum*, and *Serratia liquefaciens*). Yeasts were widely studied (34% of cases), with *Aureobasidium pullulans* (8%), *Candida* spp. (8%, especially *C. sake*), *Metschnikowia* spp. (5%, especially *M. pulcherrima*), and *Hanseniaspora uvarum* (5%) more relevant than other species, which accounted for under 2% each (i.e. *Kloeckera apiculate*, *Saccharomyces cerevisiae*, *Starmerella bacillaris*, *Pichia guilliermondii*, *P. anomala*, *P. kluyveri*, *Issatchenkia terricola*, *Lachancea thermotolerans*, and *Rhodotorula glutinis*). Filamentous fungi were studied in 14% of papers, with *Trichoderma* (*T. harzianum* and *T.*

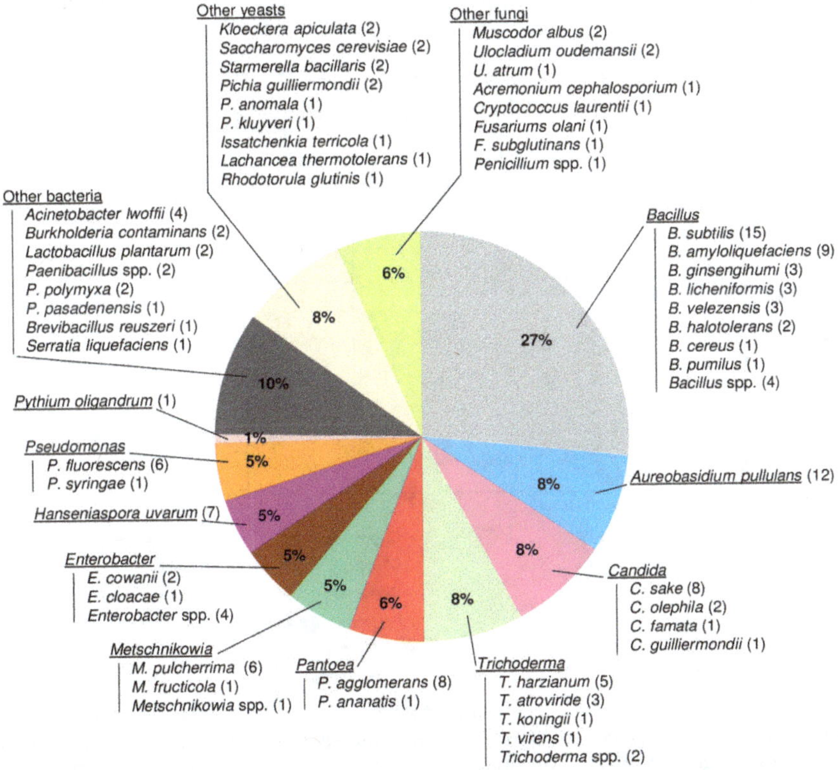

Figure 8 Representation of the microorganisms (singleton species or genera) for which papers on the *in planta* biocontrol efficacy against grey mould have been found through a systematic literature review. Sectors show the amount of papers as a percentage of total papers, while the number of papers for each microorganism is reported in brackets.

© Burleigh Dodds Science Publishing Limited, 2025. All rights reserved.

atroviride) the most represented (8%). Only one or two papers were found for the genus *Penicillium* and the singleton species *Muscodor albus*, *Ulocladium oudemansii*, *U. atrum*, *Acremonium cephalosporium*, *Cryptococcus laurentii*, *Fusariums olani*, and *F. subglutinans*.

Figure 9 shows a boxplot for BCA efficacy calculated using disease incidence and severity data, with reference to an untreated control. Regarding DM and PM, there was huge variability in the efficacy based on both microorganisms and experimental conditions.

Considering GM on leaves, *B. subtilis* showed an 88% (± 11%; *n* = 2) and 75% (± 9.8%; *n* = 6) reduction in disease incidence and severity, respectively, while *B. amyloliquefaciens* (64%; *n* = 1) and other *Bacillus* species (45.5 ± 7.5%; *n* = 11) showed lower disease severity control. The study by Ronseaux et al. (2013) showed complete suppression of the disease with *U. atrum*. High control levels, in terms of GM severity on leaves, were observed for *A. lwoffii* (74.7 ± 12.3%; *n* = 3) and *Pseudomonas* spp. (79.8 ± 12.3%; *n* = 3). By contrast, *B. reuszeri* (*n* = 4), *Enterobacter* spp. (*n* = 4), *Paenibacillus* spp. (*n* = 3), *Pantoea*

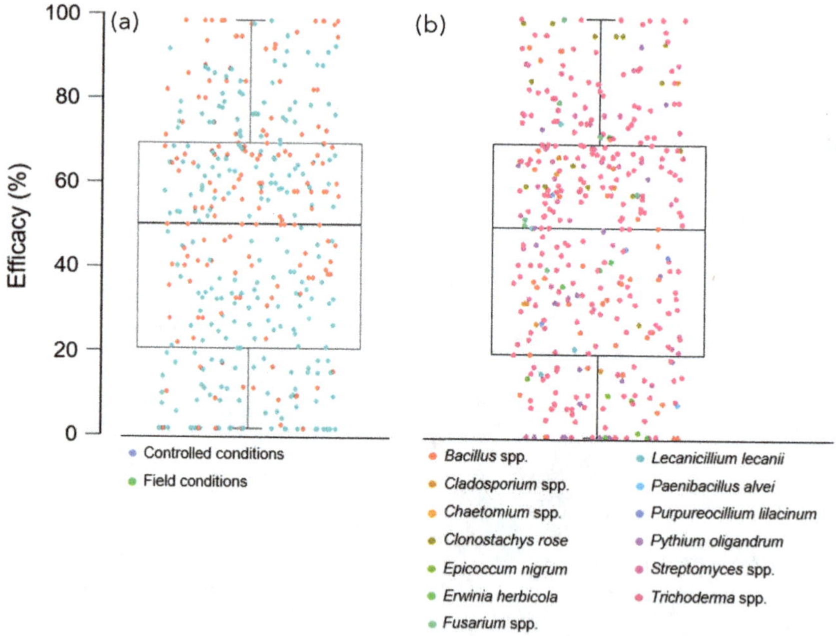

Figure 9 Box plots showing the efficacy (%) data on *in planta* grey mould biocontrol extracted from the papers shown in Figure 8. In panels A and B, disease incidence and severity data, respectively, are grouped based on experimental conditions; in panels C and D they are based on microorganisms. Efficacy was calculated using disease incidence or severity data in relation to an untreated control. The box extends from the 25th to 75th quartile of the data distribution, and the line crossing the box represents the median; whiskers extend to the maximum and minimum.

© Burleigh Dodds Science Publishing Limited, 2025. All rights reserved.

spp. (*n* = 5), and *S. liquefaciens* (*n* = 4) were studied but showed efficacies lower than 50%.

The majority of studies looked at the effect of BCAs on GM on clusters under controlled conditions. The highest control of disease incidence was observed for *C. laurentii* (*n* = 1), *I. terricola* (*n* = 3), *K. apiculata* (*n* = 2), *M. albus* (*n* = 3), and *Penicillium* spp. (*n* = 3), with efficacy greater than 80%. For *C. laurentii*, *I. terricola*, *K. apiculata*, and *Penicillium* spp., disease severity was not evaluated, while *M. albus* (*n* = 2) showed a reduction in disease severity by 97.5% (± 0.5%; *n* = 2). *H. uvarum* reduced both incidence and severity by 73.8% (±7.5%; *n* = 9) and 57.4% (± 11.0%; *n* = 5), respectively. Marsico et al. (2021) and Fedele et al. (2020b) reported reductions in disease severity as high as 100% for *L. thermotolerans* (*n* = 2), *P. oligandrum* (*n* = 2), and *Trichoderma* spp. (*n* = 2). Although disease incidence was not significantly reduced (65±10%; *n* = 2), good efficacy in terms of disease severity was calculated for *L. plantarum* (85%; *n* = 1). Biocontrol efficacy for GM incidence and/or severity ranged from 40% to 70% for *A. cephalosporium* (*n* = /1), *A. pullulans* (*n* = 3/4), *B. contaminans* (*n* = 1/1), *Candida* spp. (n = 4/4), *Enterobacter* spp. (*n* = 5), *Metschnikowia* spp. (*n* = 6/10), *Pantoea* spp. (*n* = /4), *Paenibacillus* spp. (*n* = /3), *Pichia* spp. (*n* = 3/4), and *S. bacillaris* (*n* = 1/26), with less than 30% control observed for *Pseduomonas* spp. (*n* =1/9), *R. glutinis* (*n* = 1/), and *S. cerevisiae* (*n* = 1/3).

Just 14 genera or singleton species were tested under field conditions. The best performances in terms of disease incidence reduction were recorded for *B. subtilis* (59 ± 11.3%; *n* = 3) and *Fusarium* spp. (66.6 ± 10.0%; *n* = 16), followed by *C. sake* (52.2 ± 3.9%; *n* = 18). *Candida sake* confirmed good performances in terms of disease severity, with an average efficacy of 70.8% (± 5.0%; *n* = 19). The ability of *Bacillus* species (*n* = 12) to reduce disease severity, however, was limited, with average efficacies ranging from 26 to 30%. *Trichoderma* spp. (*n* = 14) and *Ulocladium* spp. (*n* = 2) provided an average reduction of disease incidence of about 40%, with higher efficacy for disease severity: 53.7% (± 6.3%; *n* = 14) and 78% (*n* = 1), respectively. Despite less than 30% incidence control being observed for *A. pullulans* (*n* = 28) and *Metschnikowia* spp. (*n* = 1), their efficacy in reducing disease severity was 51.5% (± 8.9%; *n* = 10) and 67% (*n* = 1), respectively. The intermediate GM control observed under controlled conditions for *Enterobacter* spp., *Pantoea* spp., and *Paenibacillus* spp. was reduced in the field, with average efficacy (as disease severity) ranging from 32% to 37%. Low levels of biocontrol were confirmed for *Pseudomonas* spp., with an average efficacy of 11% (± 4.9%; *n* = 10).

7 Biocontrol efficacy against trunk diseases

A total of 27 papers on the biocontrol against the fungi causing trunk diseases were identified, contributing 429 efficacy data points. Whilst 15 studies referred

© Burleigh Dodds Science Publishing Limited, 2025. All rights reserved.

to experiments conducted under controlled conditions, 21 papers focused on field settings, with just one studying both conditions.

A total of 28 microorganisms (oomycetes, fungi, and bacteria) were considered (Figure 10). Ascomycetous fungi were prevalent (78% of papers), with the genus *Trichoderma* (66%; mainly *T. atroviride*) being prevalent over *Fusarium* (5%; mainly *F. lateritium*), and other fungi (8%; with fungal taxa considered in just one paper). Bacteria accounted for 18% of the microorganisms, with *Bacillus* the most frequently considered (13%, mainly *B. subtilis*), followed by *Streptomyces* (3%) and others (3%). *Pythium oligandrum* was the only reported oomycete.

In both controlled and field conditions, BCAs were applied to evaluate their ability to protect detached wood pieces (in controlled conditions; e.g. Silva-Valderrama et al., 2021), wounded nursery cuttings (e.g. Di Marco and Osti, 2007), grafting points (e.g. Pertot et al., 2016), or winter pruning wounds (e.g. Brown, et al., 2021) from tissue colonization by the fungi causing GTDs. In these trials, causal fungi were either artificially inoculated in the form of mycelia (e.g. Rezgui et al., 2016) or spores (e.g. Mutawila et al., 2015), or the inoculation occurred via natural spread in affected vineyards (e.g. Kotze et al., 2011). The incidence of GTDs was determined through visual assessment of symptoms such as discolouration (e.g. Di Marco et al., 2022) and wood cankers (e.g.

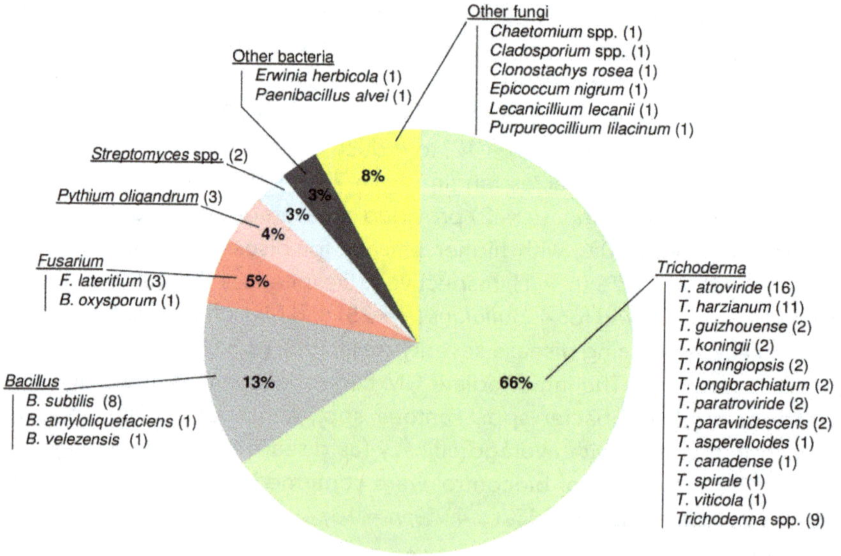

Figure 10 Representation of the microorganisms (singleton species or genera) for which papers on the *in planta* biocontrol efficacy against grape trunk diseases have been found through a systematic literature review. Sectors show the amount of papers as a percentage of total papers, while the number of papers for each microorganism is reported in brackets.

© Burleigh Dodds Science Publishing Limited, 2025. All rights reserved.

Rezgui et al., 2016), as well as by re-isolation of the inoculated fungi (e.g. John et al., 2005) and PCR analysis (e.g. Reis et al., 2021).

Figure 11 illustrates the efficacy of BCAs calculated using disease incidence data relative to an untreated control. For black foot disease, no microorganism tested achieved more than 40% efficacy under either controlled or field conditions. In the field, the most effective BCAs were *Streptomyces* (39.5 ± 8.0% efficacy, n = 18), followed by *T. atroviridae* (31.0 ± 10.5% efficacy, n = 15). For Botryosphaeria dieback, five of the 11 BACs tested under controlled conditions had more than 50% efficacy, with *C. rosea* (79.3 ± 6.8% efficacy, *n* = 12) and *T. harzianum* (74.4% ± 5.3% efficacy, *n* = 10) outperforming the other microorganisms. *Clonostachys rosea* was not tested in the field, and the efficacy of *T. harzianum* was lower (43.2 ± 6.9%, *n* = 12). Other *Trichoderma* spp. showed more than 50% efficacy in the field (*n* = 37), while *Bacillus* spp. reduced the disease by 37.5% (± 7.8% efficacy, *n* = 5). For the Esca complex, BCAs exhibited substantial variation among different taxa of microorganisms, especially under controlled conditions, ranging from 8% efficacy for *Paenibacillus alvei* (*n* = 1) to 78.3% (± 9.6%, *n* = 12) for *T. atroviridae* and 72% (± 23.0%, *n* = 2) for *T. harzianum*. *Clonostachys rosea* also demonstrated high efficacy (74.5 ± 8.3% efficacy, *n* = 6). The efficacy was lower overall for the BCAs tested in the field (average of 53.1 ± 4.4%, *n* = 29) compared to controlled conditions (average of 38.1 ± 0.7%, *n* = 111), with *Trichoderma* spp. (average of 36.3 ± 0.4%, *n* =

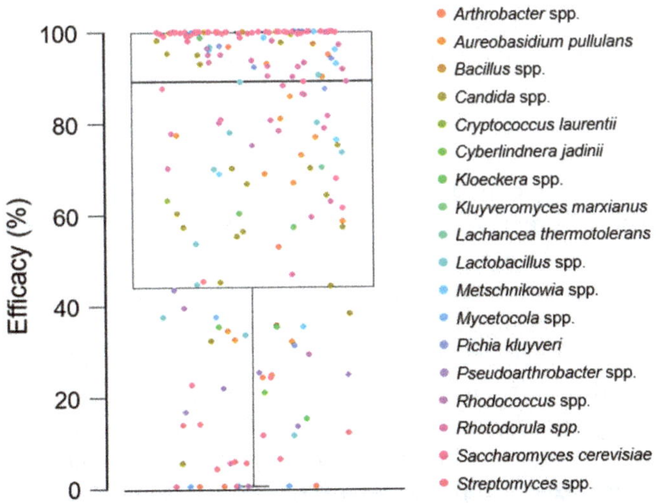

Figure 11 Box plots showing the efficacy (%) data on *in planta* biocontrol of grape trunk diseases extracted from the papers shown in Figure 10. In panel A, data are grouped based on experimental conditions; in panel B they are based on microorganisms. Efficacy was calculated using disease severity data in relation to an untreated control. The box extends from the 25th to 75th quartile of the data distribution, and the line crossing the box represents the median; whiskers extend to the maximum and minimum.

© Burleigh Dodds Science Publishing Limited, 2025. All rights reserved.

70), *Streptomyces* spp. (47.2 ± 5.9%, n = 24) and *Bacillus* spp. (45.2 ± 9.0%, *n* = 5) exhibiting higher efficacy than *P. oligandrum* (27.2 ± 9.3%, *n* = 12). For Eutypa dieback, experiments under controlled conditions ranged from no efficacy (Lecanicillium lecanii, *n* = 1) to complete (*Fusarium* spp., *n* = 2), with the remainder ranging from 32% (± 3.7%, *n* = 7, for *Bacillus* spp.) to 58.3% (± 8.7%, *n* = 3, for *T. harzianum*). When tested in the field, all BCAs showed more than 50% (*T. harzianum*, *T. atroviridae*, *Bacillus* spp.) or 60% (other *Trichoderma* and *Fusarium* spp.) efficacy.

8 Biocontrol efficacy against ochratoxigenic fungi and sour rot

A total of 17 studies evaluated the biocontrol efficacy against OF, providing 223 efficacy figures. Most studies were conducted under controlled environmental conditions (15 of 17) with the remainder in the field (De Felice et al., 2008; Dimakopoulou et al., 2008).

Under controlled conditions, BCAs and *Aspergillus* spp. were co-cultured on synthetic media (e.g. Zhu et al., 2015), or on berries (e.g. Lappa et al., 2018) or were applied to media previously contaminated with the fungus (e.g. El Khoury et al., 2017). The OTA concentration in the medium was quantified with liquid chromatography (HPLC). In the field studies (De Felice et al., 2008; Dimakopoulou et al., 2008), a cell suspension of *A. pullulans* was applied at pre-bunch closure, early veraison, and/or before harvest, and the OTA contamination produced by natural infection by *Aspergillus* spp. was determined at harvest through HPLC.

A total of 33 microorganisms from 18 genera were considered, with 30% being bacteria and 70% yeast. No filamentous fungi were tested as BCAs against ochratoxigenic fungi. Most tested BCAs belonged to *Candida* (18%, comprising six different species), followed by 18% focused on "other yeasts" (Figure 12), and 12% on *Saccharomyces cerevisiae*. Among bacteria, 18% were represented by *Streptomyces* spp., *Lactobacillus*, and *Bacillus* species, while "other bacteria" made up 12%.

Figure 13 displays the boxplot of BCA efficacy calculated using OTA concentration data with reference to an untreated control. The highest efficacy (>80%) in terms of reducing ochratoxin A (OTA) production by *A. carbonarius* was observed for the bacteria *Bacillus* (*n* = 7) and *Rhodococcus* (*n* = 3), with OTA reductions exceeding 90%. Among yeasts, *S. cerevisiae* (*n* = 43) and *Kluyveromyces marxianus* (*n* = 1) also showed more than 90% efficacy in reducing OTA production by *A. carbonarius*. Conversely, the lowest efficacy was observed for *Arthrobacter* and *Mycetocola*, with 8.2% (± 8.1; *n* = 3) and 12.4% (± 12.4; *n* = 3) efficacy, respectively, in controlling OTA production by *A. carbonarius*. *Bacillus* also exhibited complete efficacy (100 ± 0%; *n* = 3) in

© Burleigh Dodds Science Publishing Limited, 2025. All rights reserved.

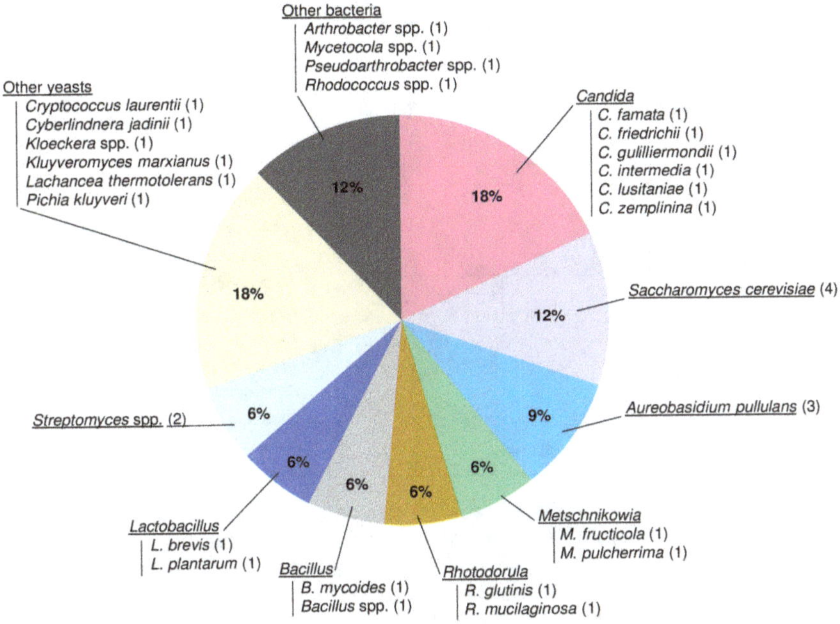

Figure 12 Representation of the microorganisms (singleton species or genera) for which papers on the biocontrol efficacy against the production of ochratoxin A (OTA) by *Aspergilla* have been found through a systematic literature review. Sectors show the amount of papers as a percentage of total papers, while the number of papers for each microorganisms is reported in brackets.

controlling OTA production by *A. niger*. In addition, *Candida* (*n* = 4), *Pichia kluyveri* (*n* = 4), and *S. cerevisiae* (*n* = 10) showed promising results (>88%) in reducing OTA production by *A. ochraceus*.

8.1 Biocontrol efficacy against sour rot (SR)

Only one paper evaluated BCA efficacy in controlling SR. Specifically, Calvo-Garrido et al. (2013) investigated the efficacy of the yeast *Candida sake* under field conditions on both disease incidence and severity on clusters. The study found that SR incidence at harvest was reduced by 27.3% and severity by 66.6%.

9 General lessons from the literature on efficacy

The biocontrol of grape diseases has been studied extensively, as demonstrated by the large number of studies and the variety of microorganisms tested. These microorganisms belong to different kingdoms and have been evaluated for their ability to control one or multiple target pathogens. They display various MoAs

© Burleigh Dodds Science Publishing Limited, 2025. All rights reserved.

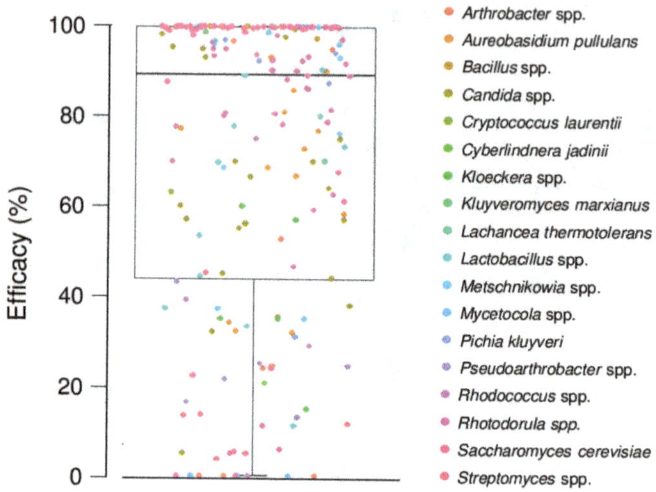

Figure 13 Box plots showing the efficacy (%) data on biocontrol against the production of ochratoxin A (OTA) by *Aspergilla* provided by different microorganisms; data were extracted from the papers shown in Figure 12. Efficacy was calculated using disease OTA concentration in relation to an untreated control. The box extends from the 25th to 75th quartile of the data distribution, and the line crossing the box represents the median; whiskers extend to the maximum and minimum.

and have diverse nutritional and environmental requirements. Only some of these studies, however, provide meaningful information on the efficacy of BCA application on plants (either under controlled conditions or in the vineyard).

The significant differences (clearly shown by the boxplots in Figures 5, 7, 9, and 11) among the experiments carried out at various locations (across continents in both the southern and northern hemispheres) and in different years, in which the disease pressure ranged from mild to extremely severe, and the environmental conditions were highly diverse, made synthesis of the results difficult. Variations in experimental settings (e.g. controlled vs field conditions, pre- vs post-infection timing, distribution methods, etc.) and methods for assessing BCA efficacy also increased this difficulty. In addition, some BCAs have been classified at the genus level only, or when the classification is provided for the species, so no information is available for the specific strain. Even when the strain details are supplied, they are rarely deposited in a mycotheque, making replication by other researchers problematic.

As a general trend, there is a clear reduction in BCA efficacy when moving from controlled environments to field studies, suggesting that the former studies primarily provide information about the potential effect of microorganisms, which must then be confirmed through repeated field experiments. Unfortunately, this secondary check is often not available. For instance, *L. thermotolerans* provided full protection against *B. cinerea* on berries

© Burleigh Dodds Science Publishing Limited, 2025. All rights reserved.

in controlled conditions, but, to the best of our knowledge, the microorganism has not been tested in the vineyard. Similarly, much of the information available on ochratoxigenic fungi comes from lab experiments only, which may limit the generalizability of these findings.

Only a few of the microorganisms tested in the aforementioned studies have become commercial plant protection products. As an example, 85 of the 176 studies included in this analysis were conducted in Europe, taking into consideration 61 microorganisms. However, only 14 of these microorganisms have been authorized to be used in viticulture in Europe, namely *Ampelomyces quisqualis* AQ10 against powdery mildew; *Bacillus pumilus* QST2808 against powdery mildew; *B. amyloliquefaciens* MBI600 against powdery mildew, grey mould, and sour rot; FZB24 against grey mould and sour rot; *Aureobasidium pullulans*, *B. amyloliquefaciens* subsp. *plantarum* D747, *B. subtilis* QST713, *Metschnikowia fructicola* NRRLY-27328, *Saccaromyces cerviasiae* LASO2, *Pythium oligandrum* M1 against grey mould; *Trichoderma asperellum* ICC012, *T. gamsii* ICC080 against grey mould, sour rot, and trunk diseases; *T. atroviride* I-1237 against trunk diseases; and *T. atroviride* SC1 against grey mould and trunk diseases. No biocontrol products are currently authorized for downy mildew or black rot.

Since robust evidence on biocontrol efficacy is essential for both the authorization of BCA-based plant protection products and their adoption by farmers, a different approach to the study of BCAs is urgently required. The following sections outline two potential approaches: (i) using meta-analysis to better utilize literature data and draw robust conclusions about biocontrol efficacy; and (ii) developing BCA-based disease control tactics based on studying the interactions between the microorganism, the status of the pathogen, the plant, and the environmental conditions.

10 Understanding and improving biological control efficacy through network meta-analysis: powdery mildew, trunk diseases, and mycotoxigenic fungi

Previous discussions have highlighted the need to critically and quantitatively evaluate the factors influencing the success or failure of BCAs in disease management. Network meta-analysis can enable the integration and analysis of results obtained from different sources of information to determine the overall effectiveness of biocontrol in relation to biocontrol agents and application-oriented factors (e.g. number of applications). Meta-analysis operates on the principle that overall knowledge is based on combining and synthesizing results from individual studies and observations (Madden and Paul, 2011; Borenstein et al., 2009; Cooper et al., 2009; Hunter and Schmidt, 2004).

© Burleigh Dodds Science Publishing Limited, 2025. All rights reserved.

In this chapter, three case studies were selected and investigated through network meta-analysis – namely powdery mildew, trunk diseases, and mycotoxigenic fungi. The case studies were selected because they impact different organs, have simple to complex aetiology, occur at varying points of the grape-growing season, and present different challenges with regard to fungicide availability for disease control.

Network meta-analyses were carried out following the method described by González-Domínguez et al. (2019). Briefly, literature records were used to extract data on BCA treatments, chemical treatments, and untreated controls. The response variables considered in the statistical analysis included: (i) disease severity on both leaves and clusters for PM; (ii) disease incidence on pruning wounds for TDs; and (iii) mycotoxin (OTA) concentration (expressed in g/kg). Disease severity and incidence, as well as OTA concentration, were directly retrieved from tables and figures in the selected papers as described above. To be included in the meta-analysis, studies had to present data on the within-study variance, as indicated in González-Domínguez et al. (2019). The result of an individual study included in a meta-analysis is generally referred to as the estimated effect size (Madden and Paul, 2011).

The meta-analysis was conducted using R software (v. 4.1.1; package "metafor"; Viechtbauer et al., 2010) by applying a multivariate random effects model to log-transformed values of disease severity, disease incidence, or OTA concentration. To verify that the variability in the estimated effect was due to heterogeneity (i.e. variance between studies), residual heterogeneity and consistency were assessed (Donegan et al., 2013; Higgins and Thompson, 2002; Jackson et al., 2012). The percentage reduction in disease due to treatment was calculated relative to the untreated control (Gonzalez-Dominguez et al., 2019; Ngugi et al., 2011), and differences in disease control were evaluated in pairwise combinations (Gonzalez-Dominguez et al., 2019).

10.1 Biocontrol reduces powdery mildew compared to untreated controls but is less effective than chemical treatments, and efficacy increases with the number of applications

Data on PM severity on leaves and clusters observed in untreated and BCA-treated plots were extracted from 64 individual field studies documented in nine papers, in which *Ampelomyces quisqualis*, *Bacillus* spp. (mainly *B. subtilis*, 94% of studies with *Bacillus*), *Pseudomonas fluorescens*, and *Trichoderma* spp. were used. In 55 of the 64 studies, control provided by BCAs was compared to both untreated and chemical control (mainly with sulphur and/or synthetic fungicides like azoxystrobin, flusilazole, and triadimefon). PM epidemics varied widely across the studies, with an average disease severity of 37.8% (a range

© Burleigh Dodds Science Publishing Limited, 2025. All rights reserved.

of 4.4% to 82.1% on leaves) and 36.7% (a range of 10.6% to 50.4% on clusters) in the untreated control. Disease severity also showed high variability among plots treated with BCAs, with an average disease severity of 20% (range 1.3% to 60%) and 20.2% (range 3.1% to 35.9%) on leaves and clusters, respectively. Lower disease severity was observed in fungicide-treated plots, with an average disease severity of 8.7% (a range of 0% to 35%) and 7.7% (a range of 0% to 29.4%) on leaves and clusters, respectively (Figure 14).

The meta-analysis facilitated a significant and robust comparison of PM control (Figure 15). The severity of PM was found to be significantly lower in plots treated with BCAs or fungicides than in untreated plots, and higher disease control was observed for fungicides than for BCAs. The average estimated efficacy was 48.7% (confidence interval, CI, 44.8 – 52.3%; n = 200) for BCAs (as a category including all the microorganisms tested) and 74.8% (CI 71.4 – 77.8%; n = 96) for the chemical control. All BCAs significantly reduced PM severity compared to the untreated control. *A. quisqualis* showed the highest variability in estimated efficacy (41.7%, CI; 12.3 – 61.3%; n = 6) due to the low number of observations. A similar efficacy was estimated for *Bacillus* spp. and *Trichoderma* spp. (45.8%, CI 41.2 – 50.0%, n = 54 and 44.6%, CI 37.6 – 50.8%, n = 104, respectively). The highest average efficacy (54.1%, CI 50.2 – 57.7%, n = 36) was estimated for *P. fluorescens*. Pairwise comparison showed that estimated effects for *A. quisqualis*, *Bacillus* spp., and *Trichoderma* spp. were not significantly different ($P > 0.7$) to

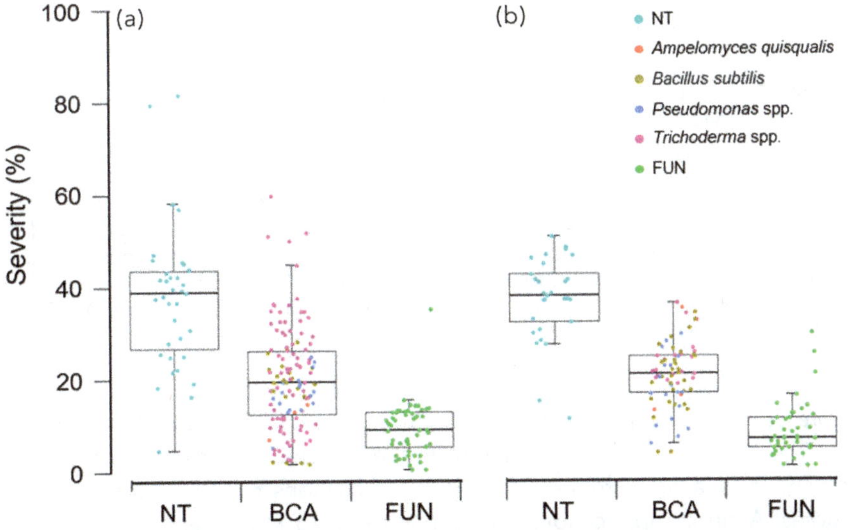

Figure 14 Powdery mildew severity (%) on leaves (A) and bunches (B) observed in field plots untreated (NT), treated with a biological control agent (BCA), or a chemical fungicide (FUN). The box extends from the 25th to 75th quartile of the data distribution, and the line crossing the box represents the median; whiskers extend to the maximum and minimum. Dot colours represent different microorganisms.

© Burleigh Dodds Science Publishing Limited, 2025. All rights reserved.

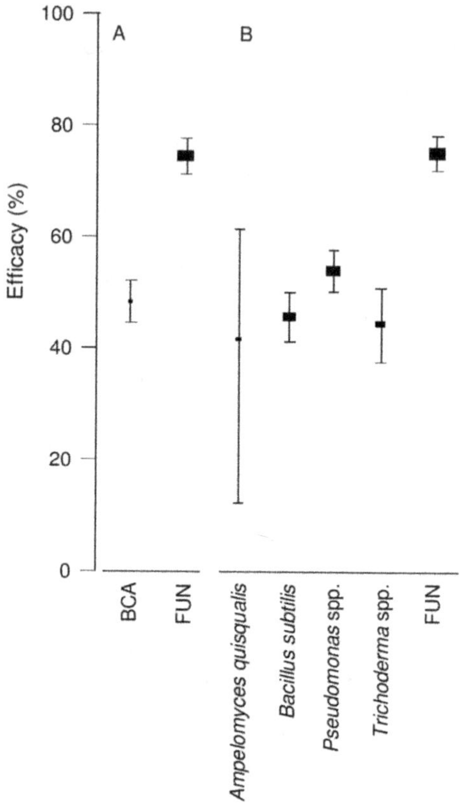

Figure 15 Forest plots showing the efficacy (%) of different treatments for the control of powdery mildew, as estimated through meta-analysis. Efficacy was calculated using disease severity data in relation to an untreated control, which is represented by the x-axis (i.e. no efficacy). Dots represent the average efficacy of treatments, and dot size increases with the precision of estimates; whiskers show a 95% confidence interval. The robustness of meta-analysis was estimated as residual heterogeneity (Q_E), degree of freedom (df), *p*-value (*P*), and % of total variation in treatment effect estimates due to heterogeneity between studies (I^2). In panel A, biological control agents (BCA) vs fungicides (FUN): QE = 3,552,924; df = 357; $P < 0.0001$; I^2 >99%. In panel B, single microorganisms vs FUN: QE = 3,543,447; df = 357; $P < 0.0001$; I^2 >95%.

one another, but all were different from *P. fluorescens* ($P < 0.05$), indicating that *P. fluorescens* provided superior PM control compared to the other BCAs.

The number of BCA applications in the selected studies varied from one to five per season, with seven or eight applications tested in two of the 64 studies. A single application significantly reduced PM severity compared to the untreated control. No significant ($P = 0.08$) differences in PM severity were observed between one and two applications, while a significant ($P < 0.001$) increase in disease control was noted for each additional application (data not shown).

© Burleigh Dodds Science Publishing Limited, 2025. All rights reserved.

10.2 Microorganisms exhibit varying levels of control over Botryosphaeria dieback, Black foot, Eutypa dieback, and Esca complex when compared to untreated plots and chemicals treatments

Data on the incidence of trunk diseases observed in untreated and BCA-treated plots were extracted from 91 individual field studies reported in 16 papers, in which *Bacillus* spp., *Fusarium* spp., *P. oligandrum*, *Streptomyces* spp., and *Trichoderma* spp. were evaluated. In 52 of the 91 studies, control provided by BCAs was compared to fungicides (mostly benomyl, in 71% of studies). Significant variability was observed in TD incidence among the studies, with an average disease incidence of 25.1% (a range of 0.5% to 92%) in the untreated control. Disease incidence also showed high variability among plots treated with BCAs and fungicides, with an average disease incidence of 14.1% (a range of 0% to 74%) and 22.5% (a range of 0% to 85.9%), respectively (Figure 16).

The meta-analysis results were statistically robust (see Figure 17). The incidence of TDs (as a whole) was significantly ($P<0.0001$) lower in plots treated with BCAs or fungicides than in untreated plots. The estimated efficacy of BCAs and fungicides averaged 40.4% (CI 32.7% – 47.2%; n = 221) and 55.3% (CI 45.8% – 63.1%; n = 121), respectively (Figure 17A). When the efficacy

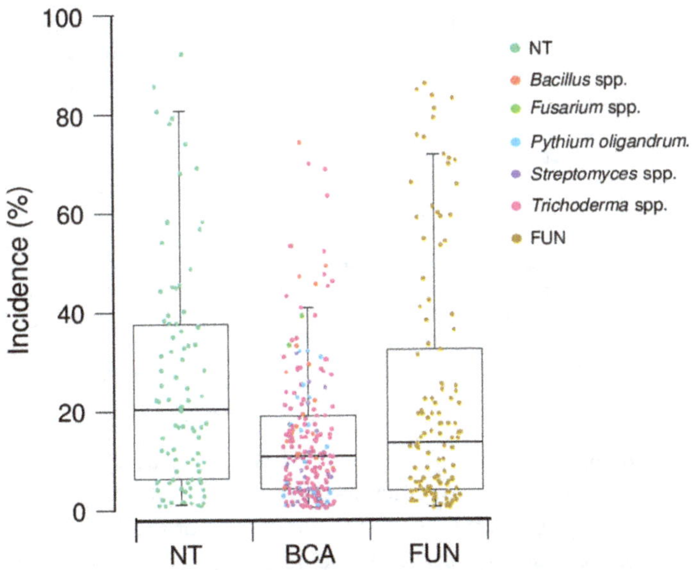

Figure 16 Incidence (%) of grape trunk diseases observed in field plots untreated (NT), treated with a biological control agent (BCA), or a chemical fungicide (FUN). The box extends from the 25th to 75th quartile of the data distribution, and the line crossing the box represents the median; whiskers extend to the maximum and minimum. Dot colours represent different microorganisms.

© Burleigh Dodds Science Publishing Limited, 2025. All rights reserved.

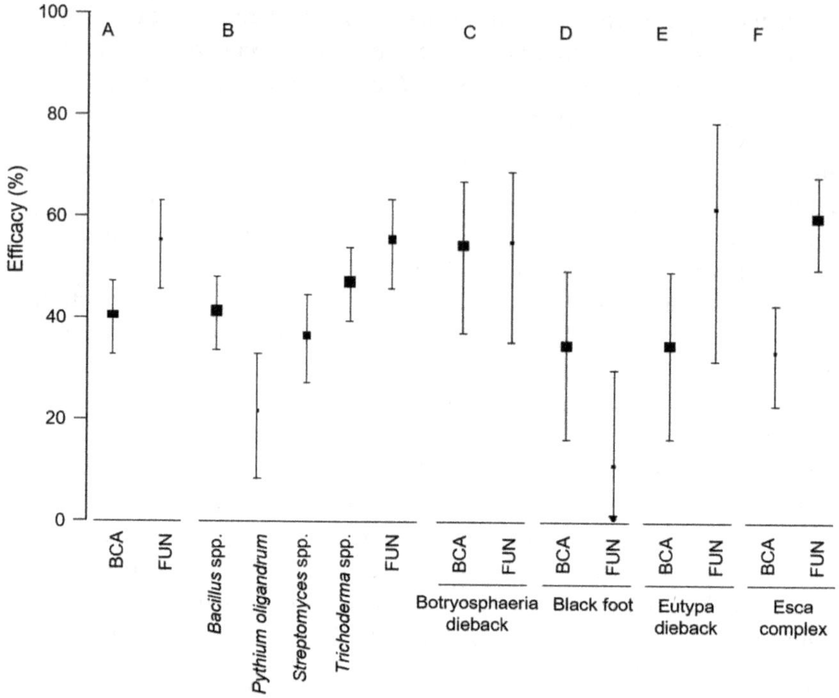

Figure 17 Forest plots showing the efficacy (%) of different treatments for the control of grape trunk diseases, as estimated through meta-analysis. Efficacy was calculated using disease severity data in relation to an untreated control, which is represented by the x-axis (i.e. no efficacy). Dots represent the average efficacy of treatments, and dot size increases with the precision of estimates; whiskers show a 95% confidence interval. Robustness of meta-analysis was estimated as residual heterogeneity (Q_E), degree of freedom (df), p-value (*P*), and % of total variation in treatment effect estimates due to heterogeneity between studies (I^2). In panel A, biological control agents (BCA) vs fungicides (FUN): QE = 60,352; df = 430; $P < 0.0001$; $I^2 > 95\%$. In panel B, single microorganisms vs FUN: QE = 59,810; df = 426; $P < 0.0001$; $I^2 > 79\%$. In panel C, BCAs vs FUN against Botryosphaeria dieback: QE = 4,225; df = 109; $P < 0.0001$; $I^2 > 80\%$. In panel D, BCAs vs FUN against Black foot disease: QE = 2,183; df = 79; $P < 0.0001$; $I^2 > 80\%$. In panel E, BCAs vs FUN against Eutypa dieback: QE = 20,208; df = 72; $P < 0.0001$; $I^2 > 80\%$. In panel F, BCA vs FUN against Esca complex: QE = 11,555; df = 176; $P < 0.0001$; $I^2 > 80\%$.

against TDs (as a whole) of individual BCAs was investigated (Figure 17B), *Bacillus* spp. showed the lowest estimated efficacy, which averaged 21.6% (CI 8.4% – 32.9%; *n* = 23), followed by *P. oligandrum* (average 37.7%, CI 50.2% – 57.7%; *n* = 24), *Trichoderma* spp. (average 43.0%, CI 35.2% – 49.8%; *n* = 148), and *Streptomyces* spp. (average 48.6%, CI 41.0% – 55.3%; *n* = 24), the latter being not significantly (*P* = 0.23) different from chemical control. *Fusarium* spp. was excluded from this comparison because there were only two studies available.

© Burleigh Dodds Science Publishing Limited, 2025. All rights reserved.

For Botryosphaeria dieback (Figure 17C), the estimated efficacy of BCAs (average of 54.3%, CI 36.9% – 66.9%; n = 48) and fungicides (average of 55%, CI 35.1% – 68.7%; n = 45) were similar to the control. For Black foot (Figure 17D), the disease in fungicide-treated plots did not differ from the untreated control (P = 0.31), while disease in the BCA-treated plot was significantly (P < 0.05) different from the untreated plot, with estimated efficacy of 34.6% (CI 16.2% – 49.0%; n = 55). In the case of Eutypa dieback (Figure 17E), plots treated with fungicides or BCAs had significantly less disease than the untreated control P < 0.01), but they did not differ (P = 0.07) from each other due to the great variability in estimated efficacies (CI 16.1% – 49.0%, n = 38, and CI 31.6% – 78.6%, n = 19, for BCAs and fungicides, respectively). For Esca complex (Figure 17F), plots treated with fungicides or BCAs had lower levels of disease than the untreated control, with estimated efficacy for fungicides (average 59.9%, CI 49.6% – 68.0%, n = 49) being significantly (P < 0.001) higher than for BCAs (average 33.4%, CI 22.7% – 42.6%, n = 90).

10.3 Both bacteria and fungi reduce ochratoxin production

Data on BCA efficacy on OTA production by *Aspergillus* were extracted from 39 individual studies reported in 13 papers. In these studies, BCAs included bacteria (of genera *Arthrobacter*, *Streptomyces*, *Bacillus*, *Lactobacillus*, *Mycetocola*, *Pseudoarthrobacter*, *Rhodococcus*) and fungi (as yeasts or yeast-like fungi, i.e. *A. pullulans*, *C. laurentii*, *C. jadinii*, *K. marxianus*, *L. thermotolerans*, *Metschnikowia* spp., *Saccharomyces* spp.). OTA concentrations varied widely across the studies, from 0 to more than 2,000 µg/kg, regardless of treatment (Figure 18).

Results of the meta-analysis were robust (Figure 19). BCAs significantly (P < 0.0001) lowered the estimated OTA concentration by 71% (CI 53.8 – 81.8%; n = 104), with no differences (P > 0.12) between bacteria (69.3% efficacy, CI 58.0 – 77.6%; n = 77) and fungi (81.2%, CI 64.0 – 90.1%; n = 27).

11 Improving biocontrol efficacy through understanding interactions between pathogen, plant and environmental conditions: *Plasmopora viticola*, *Erysiphe necator*, and *Botrytis cinerea*

As shown above, meta-analysis can improve our understanding of the literature data published to date and allow robust insight into the practical use of BCAs. Unfortunately, not much of the current literature can be used for this kind of statistical analysis, because many published papers lack the necessary information. Meta-analysis cannot account for variations in the experimental protocols used across studies, including the timing of BCA applications in vineyards with respect to grape phenology, the epidemiology

© Burleigh Dodds Science Publishing Limited, 2025. All rights reserved.

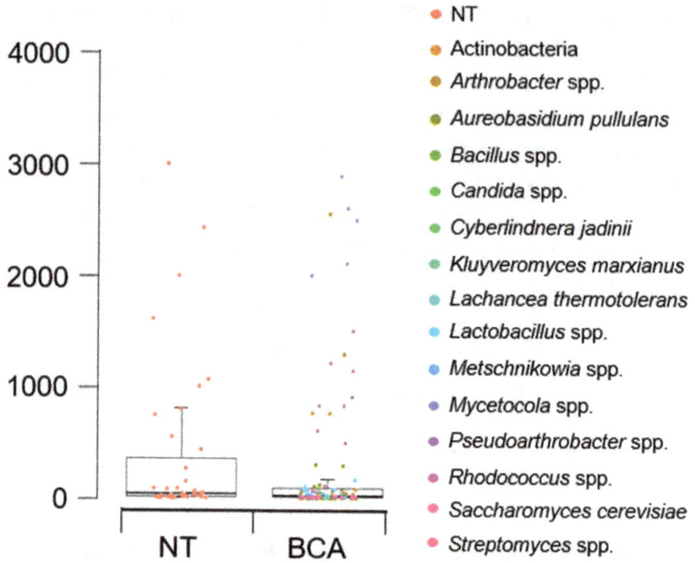

Figure 18 Concentration of ochratoxin A (OTA) (µg/kg) produced by *Aspergilla* on untreated artificial media (NT) and treated with an BCA. The box extends from the 25th to 75th quartile of the data distribution, and the line crossing the box represents the median; whiskers extend to the maximum and minimum. Dot colours represent different microorganisms.

of the target pathogen, as well as the environmental conditions that the BCA faces following its application. These factors are of paramount importance and are responsible for the huge variability that any BCA exhibits across studies, as demonstrated above. A scenario study carried out by Fedele et al. (2020d), using a mathematical model that incorporates the MoA of BCAs, demonstrated that factors such as timing with respect to the pathogen (preventative vs curative intervention), temperature, and moisture requirements for BCA growth and survival, all significantly influenced biocontrol efficacy. Temperature and moisture accounted for more than 90% of the variability in biocontrol outcomes.

Since the unpredictability of the control efficacy following a BCA application is one of the main reasons for the limited adoption of biocontrol by farmers (He et al., 2021), a different approach is necessary, where BCA applications are based on the status of the pathogen, the plant, and the environmental conditions (Fedele et al., 2020d; Altieri et al., 2023). The following paragraphs offer insights for downy and powdery mildews, and for grey mould.

11.1 Breaking the life cycle of Plasmopora viticola

Plasmopara viticola survives the winter as oospores in the leaf residue above the soil or in soil. Oospores break dormancy in the year following their formation

© Burleigh Dodds Science Publishing Limited, 2025. All rights reserved.

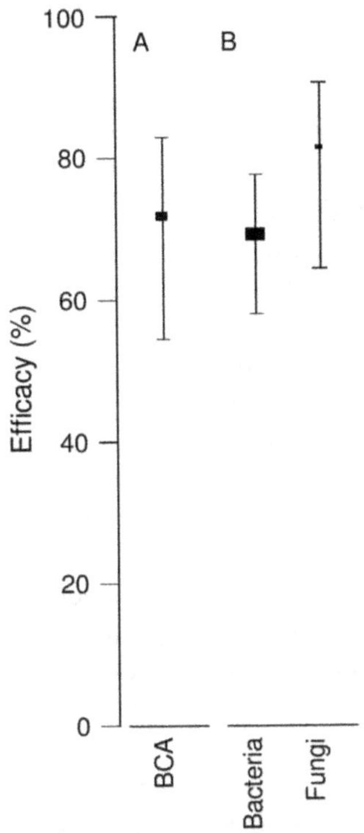

Figure 19 Forest plots showing the efficacy (%) of different treatments for the control of ochratoxin A (OTA) (μg/kg) production by *Aspergilla*, as estimated through meta-analysis. Efficacy was calculated using OTA concentration in an untreated control, which is represented by the x-axis (i.e. no efficacy). Dots represent the average efficacy of treatments, and dot size increases with the precision of estimates; whiskers show the 95% confidence interval. Robustness of meta-analysis was estimated as residual heterogeneity (Q_E), degree of freedom (df), p-value (*P*), and % of total variation in treatment effect estimates due to heterogeneity between studies (I^2). In panel A, biological control agents (BCA): QE = 215,351; df = 141; $P < 0.0001$; $I^2 > 99\%$. In panel B, biocontrol by bacteria vs yeasts: QE = 191,500; df = 141; $P < 0.0001$; $I^2 > 99\%$.

or in the next few years (Caffi et al., 2011). They repeatedly germinate in spring (typically between bud break and late July), producing sporangia that release bi-flagellated motile zoospores in water. Zoospores survive in water until they are dispersed by rain to the vegetation; otherwise, they die. Upon reaching the vegetation, zoospores swim to the stomata, encyst, and germinate, and the germ tubes penetrate the stomata opening, causing (primary) infection that results in the typical DM lesions found on leaves and clusters. Asexual sporangia are then produced on DM lesions, spread through air currents, and cause secondary infections that further progress the disease. Oospores are primarily responsible for the establishment of disease in the vineyard and for maintaining its presence throughout the season (Gessler et al, 2011). Prevention

© Burleigh Dodds Science Publishing Limited, 2025. All rights reserved.

of primary infections is key for successful DM control, currently achieved by repeated fungicide applications.

BCAs can contribute to the control of *P. viticola* through various mechanisms: (i) parasitizationu of oospores and consequent reduction of the oospore pool; (ii) killing of zoospores or interference with their normal swimming behaviour in water above the ground (i.e. before their dispersal to plants) or on the plant surfaces (i.e. before their gemination and penetration); and (iii) suppression of the production or germination of secondary sporangia, and consequent prevention of secondary infections.

Concerning (i), oospore parasitization in soils has been documented for other oomycetes, such as *Phytophthora erythroseptica*, *Ph. megasperma* var *sojae*, *Ph. cactorum*, and *Pytium* spp. by oomycetes, hyphomycetes, chytridiomycetes, actinomycetes, and bacteria (Sneh et al., 1977; Wynn and Epton, 1979; Hoch and Abawi, 1979; Sutherland and Lockwood, 1984; Khan et al., 1993) but not for *P. viticola*. Some microorganisms were able to prevent leaf infection in bioassays where oospores were kept in water and in the presence of floating leaf discs. Disease suppression on leaf discs, however, was inconsistent over repeated experiments (Dagostin et al., 2006; Vecchione et al., 2007).

Concerning (ii), a *Paenibacillus* sp. isolated from the rhizosphere of *Sorghum bicolor*, and its peptide paenimyxin, inhibited the mobility of *P. viticola* zoospores and decreased infection sites on leaf discs when added to the sporangial suspension used for artificial inoculation (Hao et al., 2017). Secondary metabolites produced by a soil-isolated *Streptomyces* sp., an endophytic *Fusarium* sp. (from *Piper guinense*), and a marine *Streptomyces* sp. also exhibited motility inhibitory activity (up to 100%) and lytic activity (up to >90%) against zoospores (Abdalla et al., 2011; Kongue Tatong et al., 2014; Dame et al., 2016). The effectiveness of these microorganisms under natural conditions has not yet been investigated; we can speculate that they could survive better in soil rather than on plant surfaces, thus inhibiting zoospores originating from macrosporangia, but this needs to be confirmed by additional research.

Concerning (iii), an *Epicoccum nigrum* strain isolated from DM-affected leaves in the field was found to encase sporangia with abundant mycelium, preventing their dispersal. Any antagonistic effect, however, was not observed and its ability to colonize leaves remains questionable (Kortekamp, 1997). A strain of *Fusarium proliferatum,* isolated from atypical DM lesions, was effective in reducing sporangial production on leaf discs by 97%. Its hyphae were seen to consistently coil around and inside sporangiophores (Falk et al., 1996). Similar results were also observed with other *Fusarium* spp. (Ghule et al., 2018). Strains of *Aletrnaria alternata* from unusual DM symptoms completely inhibited sporulation on leaf discs and caused ultrastructural changes in *P. viticola* hyphae

© Burleigh Dodds Science Publishing Limited, 2025. All rights reserved.

and haustoria, due to the production of toxic metabolites (Musetti et al., 2006). In a wide-ranging study, 47 of 1,700 organisms applied at infection time of the leaf discs, together with *P. viticola,* were able to reduce or prevent sporulation (Vecchione et al., 2004). *Acremonium* spp. isolated from asymptomatic grapevine leaves were reported as hyperparasite of *P. viticola* in leaves, with occasional formation of sporangiophores and sporangia in the leaves colonized by the endophyte, or with deformed sporangiophores (Burruano et al. 2008). These fungi also produce secondary metabolites that inhibit sporangial germination – as the ratio between germinated (empty) and non-germinated (degenerated) sporangia – with different inhibition degrees based on species and strains of the same species (Assante et al., 2005; Arnone et al., 2008, 2009; Lo Piccolo et al., 2015). Inhibition of sporangial germination was also observed for various unspecified bacteria (seven of 17 tested) (Vecchione et al., 2004). To our knowledge, the field activity of the abovementioned microorganisms has only been tested by Falk et al. (1996), who found that certain fungi reduced DM severity on clusters and leaves by 53 to 99% following weekly applications. However, none of these microorganisms have been further developed as a commercial product.

11.2 Understanding the epi-season of Erysiphe necator

The life cycle of *E. necator* spans two viticultural seasons, a concept known as the "epi-season" (Magarey and Moyer, 2010), in which the current season leaves a "legacy" for the next. Sexual fruiting bodies (the chasmothecia) form from late summer to autumn on PM colonies, especially after harvest, when the interruption of fungicidal treatments gives the pathogen the opportunity to develop. Mature chasmothecia (black in colour) are dispersed to the trunks of the vines by rainfall, where they remain trapped in the bark and remain viable until the following spring. Chasmothecia that fall to the ground typically lose their vitality during winter. Ascospores are released by chasmothecia in spring (normally between bud break and fruit set; Rossi et al., 2010). If environmental conditions are favourable, they can cause primary (or ascosporic) infections. These infections have symptoms that are atypical and often go unnoticed. However, ascosporic infections play a key role for the establishment of PM in the vineyard and for the initial multiplication of the asexual inoculum, the conidia, which is responsible for secondary infections and, therefore, the polycyclic and "explosive" nature of PM epidemics.

In the efficacy studies mentioned above, the microbial products were used in late spring and/or summer to control the disease in its epidemic phase. In this period, PM colonies produce huge quantities of conidia under a wide range of environmental conditions, which are dispersed by air currents and cause repeated infection cycles that sporulate in turn following a latent period

© Burleigh Dodds Science Publishing Limited, 2025. All rights reserved.

of five to six days. The control of PM in these conditions is challenging, and even sulphur and synthetic fungicides do not enable full disease control (Legler et al., 2011). During summer, the air temperature is high and free moisture in the grapevine canopy is low. These environmental conditions do not limit the development of *E. necator* (Gadoury et al., 2012) but may prevent sustained growth of other microorganisms, including BCAs.

Given the life cycle of *E. necator* and the concept of the epi-season, PM control should be anticipated and targeted to the sexual phase of the fungus. This means: (i) controlling chasmothecia in the development phase, between late summer and early autumn; (ii) killing mature chasmothecia on the vine bark in winter before bud break; and (iii) preventing ascosporic infections during the following spring. In epidemiological terms, this approach delays the epidemic phase of the disease (Vanderplank, 1963) in a period in which clusters have already acquired ontogenic resistance to *E. necator* (Gadoury et al., 2003).

The suppression of chasmothecia production in (i) can be achieved by using *A. quisqualis*, a hyperparasite of *E. necator* that naturally colonizes mycelia, conidia, and chasmothecia when they are young, i.e. they are yellow to orange (Kiss et al., 2004; Angeli et al., 2009; Legler et al., 2012). Parasitized chasmothecia are typically dull, fawn-coloured, and flaccid (Falk et al., 1995). In controlled conditions, *A. quisqualis* was able to suppress chasmothecia formation by 41%, while field application of *A. quisqualis* during the pre- and post-harvest delayed disease onset and epidemic development in the following season by one to four weeks, and lowered both disease incidence and severity on clusters in the following season by approximately 50% compared to an untreated control (Caffi et al. 2013; Legler et al. 2016). In an organic strategy based on fall application of *A. quisqualis* against chasmothecia and sulphur treatments targeted to prevent ascosporic infections in spring, the average disease control was >95%, and similar to the control obtained by using synthetic fungicides (Caffi et al., 2013). The trunk sanitation in (ii) is commonly obtained with lime sulphur (Gadoury et al., 1994), fungicidal oils (Thiessen et al., 2018), or Bordeaux mixture (Todaro and Miles, 2018). No BCAs are yet known to parasitize overwintered chasmothecia.

Concerning (iii), application of *Bacillus*-based BCAs against ascospric infection appears promising. These products have different MoAs (Cawoy et. al., 2011; Tran et al., 2022) that inhibit spore germination and hyphal growth. Their efficacy has been rated as good and reliable under low to medium disease pressure, with reduced efficacy under high disease pressure (Adaskaveg et al., 2022). Since the PM pressure is lower for ascospric than for conidial infections, these treatments are typically recommended between bud break and full bloom (Adaskaveg et al., 2022). It may also be considered that ascosporic and conidial infections have different needs that also suggest using BCAs against the former. Ascosporic infections are triggered by rain and require water on plant surfaces,

© Burleigh Dodds Science Publishing Limited, 2025. All rights reserved.

with a temperature range of 5–28°C (optimum 20–25°C). Conidial infections are hindered by rain and the presence of water but are favored by high relative humidity, with an optimal temperature of 20–25°C (but infection is possible up to 35°C).

11.3 Target the multiple infection pathways of **Botrytis cinerea**

The life cycle of *B. cinerea* is complex, and multiple infection pathways exist (Elmer and Michailides, 2007; González-Domınguez et al., 2015). At vine flowering, latent infections occur, which will remain latent until berry ripening, when they can result in rotten berries (infection pathways I, IIa, and IIb of Elmer and Michailides, 2007). From fruit set to veraison, *B. cinera* saprophytically colonizes bunch trash (i.e. calyptras, dead stamens, aborted flowers and berries, and tendrils), and produces conidia that may serve for conidial infection of berries (infection pathway III and IV). From veraison onwards, berries can be infected by conidia (infection pathway Va) or by the mycelium from adjacent affected berries (the berry-to-berry infection; pathway Vb; González-Domínguez et al., 2015).

This complexity requires multiple interventions during the season, specifically at flowering (known as intervention in A), pre-bunch closure (B), veraison (C), and ripening (D) to target the different infection pathways (González-Domínguez et al., 2019). Biocontrol of GM may then comply with this complexity. From pre-flowering to fruit set, BCAs may be able to prevent both conidial germination on inflorescences and the establishment of latent infection of berries. From fruit-set to pre-bunch closure, they may competitively colonize bunch trash as a saprophyte and prevent sporulation by *B. cinerea*. From veraison until harvest, BCAs may colonize the berry surface and prevent infection by conidia and aerial mycelium (berry-to-berry infection).

At the abovementioned timings, BCAs may express different MoAs and develop on different plant substrates. Intervention in A requires the capability of BCA to inhibit conidial germination through the production of bioactive metabolites and/or to induce resistance in clusters to prevent latent infections. Intervention in B implies the BCA's capability to compete with *B. cinerea* for the saprophytic growth on bunch trash and inhibit its sporulation. Interventions in C and D require the BCA to compete with the pathogen for the infection sites (i.e. microcracks and wounds on the berry skin), to inhibit conidial germination and hyphal growth, and/or to induce resistance, e.g. by increasing the production of phytoalexins in the outer berry layers. Mechanisms of action of microbial biocontrol agents against *B. cinerea* have recently been reviewed by Roca-Couso et al. (2021). To add complexity, the different plant substrates (flowers and flower parts, bunch trash, and berries at different ripening stages) display different biochemical characteristics (e.g. pH, nutrient types and availability,

© Burleigh Dodds Science Publishing Limited, 2025. All rights reserved.

water potential) that also vary during the season (Fedele et al., 2020b, 2020c). For instance, *Bacillus amyloliquefaciens* grow more profusely on berries at veraison and berry softening than at earlier or later stages; *Aureobasidium pullulans* grow more at veraison, the growth of *Metschnikowia fructicola* increases as berry ripening progresses, and *Trichoderma atroviride* show the opposite (Altieri et al., 2022).

Disease control interventions typically start at flowering and continue until pre-harvest, which means that BCAs need to be adapted to the varying weather conditions from spring to late summer. Temperature, humidity, and rainfall have been shown to influence the growth and efficacy of different BCAs in various ways (Fedele et al., 2020a, Altieri et al., 2022, 2023). Unfortunately, little information is available to date regarding the environmental requirements of the different BCAs (Fedele et al., 2020a). The environmental niche approach proposed by Fedele et al. (2020a) – which focuses on the environmental conditions necessary for the presence of a species and the maintenance of its population – could aid the selection of the appropriate BCA to be used in field conditions for a specific intervention, as well as the screening of new BCAs.

A multi-criteria approach is essential for effective biocontrol of GM, in which different BCAs are chosen based on the targeted infection pathway, the required MoA, and the environmental conditions at the time of intervention and in the days following it (Pertot et al., 2017; Fedele et al., 2020d).

12 Conclusions

Despite the substantial number of studies published on the biocontrol of grape diseases, the availability of commercial products remains limited (with obvious variations among countries) and acceptance by farmers remains low.

The reasons to why BCAs are less used than anticipated have been widely discussed (e.g. Velivelli et al., 2014; Singh et al., 2020; Lahlali et al., 2022). In a recent paper, Collinge et al. (2022) discussed what has been achieved to date and outlined the steps required for BCAs to become a standard part of the disease management toolkit, as disease resistance and pesticides are. The focus of this paper was on microorganism selection, development of a commercial product (from production to formulation), delivery systems, and regulatory and industrial approval. This chapter, however, specifically addresses BCA efficacy when applied in vineyards.

Two main limitations arise from the literature analysis: (i) BCAs typically provide lower disease control compared to chemical fungicides, particularly under high disease pressure; and (ii) there is significant variability in biocontrol efficacy across experiments, which undermines famer trust. The huge variability among microorganisms (species and strains) used, application timings and frequency, methods for efficacy assessment, and varying environmental

© Burleigh Dodds Science Publishing Limited, 2025. All rights reserved.

conditions make comparison among the literature studies difficult and inconsistent so that it is virtually impossible to create practical guidelines for BCA use against grape disease. The situation is further complicated by the lack of detailed experiment information and incomplete reporting of experimental errors, limiting the potential for formal meta-analysis. These limitations should be considered by researchers when planning experiments and writing papers.

Another notable issue from the literature analysis is that BCA application timing is often adopted from traditional fungicide scheduling, resulting in calendar-based applications during the typical disease season. This approach overlooks the fact that biocontrol relies on living microorganisms and their interaction with the pathogen and the plant (the plant site that is challenged by the pathogen), in a variable environment. Much progress is needed in this area.

An innovative, step-by-step approach is proposed here. The first step is to clearly define the pathogen stage for which biocontrol is required (e.g. oospores of *P. viticola* on the vineyard floor, developing chasmothecia of *E. necator* of the vegetation in late season, or saprophytic colonization of bunch trash to prevent growth and sporulation of *B. cinerea*). The second step is to define the MoA that is required to control the pathogen at a specific stage (e.g. parasitism of oospores or chasmothecia, or competition for the bunch trash material). The third step is to create a list of microorganisms responding to the following criteria: (i) possess the required MoA; (ii) ability to grow on the substrate of interest (e.g. leaf residues and soil where the *P. viticola* oospores survive, the leaves and clusters where the PM colonies develop chasmothecia, or the bunch trash that is colonized by *B. cinerea*); and (iii) share the same environmental niche as the pathogen at the target stage, so that the BCA may grow when the pathogen grows. The fourth step is to perform efficacy experiments across multiple environments and seasons to confirm the efficacy of the selected BCAs against the target pathogen and stage. By following this approach, positive experimental outcomes can result in practical guidelines for farmers.

13 Data availability statement

The list of published papers used in data collection and supporting the conclusions of this chapter can be made available by the authors upon request.

14 Conflict of interest

The authors declare that their research was conducted in the absence of any commercial or financial relationships that could be construed as a potential conflict of interest.

© Burleigh Dodds Science Publishing Limited, 2025. All rights reserved.

15 Author Contributions

RV and SI mainly contributed to the conception and design of the study. AV, CLM, JT, and SI contributed to data collection. AV, RV, and SI carried out the analysis of results. AV, CLM, JT, RV, and SI contributed to writing the original draft preparation. BP and FG contributed to reviewing and editing the original draft. All authors wrote the manuscript, contributed to the article, and approved the submitted version.

16 Acknowledgments

AV conducted this study within the Doctoral School on the Agro-Food System (AgriSystem) at the Università Cattolica del Sacro Cuore, Piacenza, Italy. SI was partially supported by the "National Research Centre for Agricultural Technologies (Agritech)" CN00000022. PNRR-M4C2, Investment 1.4 - Funded by the European Union - NextGenerationEU" - CUP: J33C22001160007.

17 References

Abdalla, M. A., et al. (2011). Khatmiamycin, a motility inhibitor and zoosporicide against the grapevine downy mildew pathogen *Plasmopara viticola* from Streptomyces sp. ANK313. *The Journal of Antibiotics*, *64*(10), 655–659.

Ab Rahman, S. F. S., et al. (2018). Emerging microbial biocontrol strategies for plant pathogens. *Plant Science*, *267*, 102–111.

Adaskaveg, J. E., et al. (2022). *Fungicides, bactericides, biocontrols, and natural products for deciduous tree fruit and nut, citrus, strawberry, and vine crops in California 2022*. Riverside: University of California. 74 p.

Ahmed, M. F. A. (2018). Evaluation of some biocontrol agents to control Thompson seedless grapevine powdery mildew disease. *Egyptian Journal of Biological Pest Control*, *28*, 1–7.

Altieri, V., et al. (2022). Biocontrol of *Botrytis cinerea* as influenced by grapevine growth stages and environmental conditions. *XXVII Congress of the Italian Phytopathological Society*. September 21st–23rd, 2022.

Altieri, V., et al. (2023). Efficacy of preharvest application of biocontrol agents against gray mold in grapevine. *Frontiers in Plant Science*, *14*, 1154370. https://doi.org/10.3389/fpls.2023.1154370

Angeli, D., et all. (2009). Occurrence of *Erysiphe necator* chasmothecia and their natural parasitism by *Ampelomyces quisqualis*. *Phytopathology*, *99*(6), 704–710.

Angeli, D., et al. (2012). Existence of different physiological forms within genetically diverse strains of *Ampelomyces quisqualis*. *Phytoparasitica*, *40*, 37–51.

Arnone, A., et al. (2008). Acremine G, dimeric metabolite from cultures of *Acremonium byssoides* A20. *Journal of Natural Products*, *71*(1), 146–149.

Arnone, A., et al. (2009). Acremines H–N, novel prenylated polyketide metabolites produced by a strain of *Acremonium byssoides*. *Tetrahedron*, *65*(4), 786–791.

Assante, G., et al. (2005). Acremines AF, novel secondary metabolites produced by an endophytic strain of *Acremonium* sp., isolated in grape leaves infected by *Plasmopara viticola*. *Tetrahedron*, *61*(32), 7686–7692.

© Burleigh Dodds Science Publishing Limited, 2025. All rights reserved.

Banani, H., et al. (2014). Characterization of resistance mechanisms activated by *Trichoderma harzianum* T39 and benzothiadiazole to downy mildew in different grapevine cultivars. *Plant Pathology*, *63*(2), 334–343.

Battilani, P., et al. (2006). European research on ochratoxin A in grapes and wine. *International Journal of Food Microbiology*, *111*, S2–S4.

Biesbroek, G. R., et al. (2013). On the nature of barriers to climate change adaptation. *Regional Environmental Change*, *13*, 1119–1129. https://doi.org/10.1007/s10113-013-0421-y

Bleyer, K., et al. (2023). Using ontogenetic resistance of grapevine for fungicide reduction strategies. *European Journal of Plant Pathology*, *165*(1), 115–124.

Bois, B., et al. (2017). Climate vs grapevine pests and diseases worldwide: the first results of a global survey. OENO *One*, *51*(2), 133–139.

Borenstein, M., et al. (2009). *Introduction to meta-analysis*. Chichester, UK: John Wiley & Sons.

Brown, A. A., et al. (2021). Pruning-wound protectants for trunk-disease management in California table grapes. *Crop Protection*, *141*, 105490.

Bruisson, S., et al. (2019). Endophytes and epiphytes from the grapevine leaf microbiome as potential biocontrol agents against phytopathogens. *Frontiers in Microbiology*, *10*, 2726.

Burruano, S., et al. (2008). Interaction between Acremonium byssoides and *Plasmopara viticola* in *Vitis vinifera*. *Phytopathologia Mediterranea*, *47*(2), 122–131.

Caffi, T., et al. (2011). Long-term survival of *Plasmopara viticola* oospores. *IOBC/WPRS Bulletin*, *66*, 111–114.

Caffi, T., et al. (2013). Combining sanitation and disease modelling for control of grapevine powdery mildew. *European Journal of Plant Pathology*, *135*, 817–829.

Calvo-Garrido, C., et al. (2013). Biological control of botrytis bunch rot in organic wine grapes with the yeast antagonist *Candida sake* CPA-1. *Plant Pathology*, *62*(3), 510–519.

Calvo-Garrido, C., et al. (2013). *Candida sake* CPA-1 and other biologically based products as potential control strategies to reduce sour rot of grapes. *Letters in Applied Microbiology*, *57*(4), 356–361.

Calvo-Garrido, C., et al. (2019). Microbial antagonism toward Botrytis bunch rot of grapes in multiple field tests using one *Bacillus ginsengihumi* strain and formulated biological control products. *Frontiers in Plant Science*, *10*, 105.

Candel, J. J. (2014). Food security governance: A systematic literature review. *Food Security*, *6*, 585–601.

Castoria, R., et al. (2001). *Aureobasidium pullulans* (LS-30) an antagonist of postharvest pathogens of fruits: study on its modes of action. *Postharvest Biology and Technology*, *22*, 7e17.

Cawoy, H., et al. (2011). *Bacillus*-based biological control of plant diseases. In M. Stoytcheva (Ed.), *Pesticides in the modern world-pesticides use and management* (pp. 273–302). InTech Academic Press.

Choudhary, D. K. & Johri, B. N. (2009). Interactions of *Bacillus* spp. and plants–with special reference to induced systemic resistance (ISR). *Microbiological Research*, *164*(5), 493–513.

Collinge, D. B., et al. (2022). Biological control of plant diseases –What has been achieved and what is the direction? *Plant Pathology*, *71*(5), 1024–1047.

Compant, S., et al. (2013). Use of beneficial bacteria and their secondary metabolites to control grapevine pathogen diseases. *BioControl*, *58*, 435–455.

© Burleigh Dodds Science Publishing Limited, 2025. All rights reserved.

Conway, G. R. & Pretty, J. N. (1991). *Unwelcome harvest: agriculture and pollution*. London: Earthscan Publishing.

Cooper, H. & Hedges, L. V. (2009). Research synthesis as a scientific process. In A. H. Cooper et al. (Eds.), *The handbook of research rynthesis and meta-analysis* (2nd ed., pp. 3–16). Thousand Oaks, CA: Sage Publications.

Cordero-Bueso, G., et al. (2017). Wild grape-associated yeasts as promising biocontrol agents against *Vitis vinifera* fungal pathogens. *Frontiers in Microbiology*, *8*, 2025.

Corkley, I., et al. (2022). Fungicide resistance management: Maximizing the effective life of plant protection products. *Plant Pathology*, *71*(1), 150–169.

Crisp, P., et al. (2006). An evaluation of biological and abiotic controls for grapevine powdery mildew. 1. Greenhouse studies. *Australian Journal of Grape and Wine Research*, *12*(3), 192–202.

Dagostin, S., et al. (2006). Potential use of biocontrol agents to prevent *Plasmopara viticola* oospore germination. *IOBC/WPRS Bulletin*, *29*(11), 43.

Dagostin, S., et al. (2011). Are there alternatives to copper for controlling grapevine downy mildew in organic viticulture? *Crop Protection*, *30*(7), 776–788.

Dame, Z. T., et al. (2016). Oligomycins and pamamycin homologs impair motility and induce lysis of zoospores of the grapevine downy mildew pathogen, Plasmopara viticola. *FEMS Microbiology Letters*, *363*(16), fnw167.

De Curtis, F., et al. (2012). Environmental factors affect the activity of biocontrol agents against ochratoxigenic *Aspergillus carbonarius* on wine grape. *International Journal of Food Microbiology*, *159*(1), 17–24.

De Felice, D. V., et al. (2008). Strains of *Aureobasidium pullulans* can lower ochratoxin A contamination in wine grapes. *Phytopathology*, *98*(12), 1261–1270.

Dimakopoulou, M., et al. (2008). Phyllosphere grapevine yeast *Aureobasidium pullulans* reduces *Aspergillus carbonarius* (sour rot) incidence in wine-producing vineyards in Greece. *Biological control*, *46*(2), 158–165.

Di Marco, S. & Osti, F. (2007). Applications of Trichoderma to prevent *Phaeomoniella chlamydospora* infections in organic nurseries. *Phytopathologia Mediterranea*, *46*(1), 73–83.

Di Marco, S., et al. (2022). Activity of *Trichoderma asperellum* strain ICC 012 and *Trichoderma gamsii* strain ICC 080 toward diseases of esca complex and associated pathogens. *Frontiers in Microbiology*, *12*, 813410.

Donegan, S., et al. (2013). Assessing key assumptions of network meta-analysis: a review of methods. *Research Synthesis Methods*, *4*(4), 291–323.

El Khoury, R., et al. (2017). Ability of soil isolated actinobacterial strains to prevent, bind and biodegrade ochratoxin A. *Toxins*, *9*(7), 222.

Elad, Y. & Freeman, S. (2002). Biological control of fungal plant pathogens. In *Agricultural applications* (pp. 93–109). Berlin, Heidelberg: Springer.

Elmer, P. A. G. & Michailides, T. J. (2007). Epidemiology of *botrytis cinerea* in orchard and vine crops. In Y. Elad, et al. (Eds.), *Botrytis: Biology, pathology and control* (pp. 243–272). Dordrecht: Springer Netherlands. https://doi.org/10.1007/978-1-4020-2626-3_14

Elmer, P. A. G. & Reglinski, T. (2006). Biosuppression of *Botrytis cinerea* in grapes. *Plant Pathology*, *55*(2), 155–177.

European Commission. (2020). *A farm to fork strategy for a fair, healthy and environmentally-friendly food system*. COM(2020) 381 Final. Brussels: European Commission. https://eur-lex.europa.eu/legal-content/EN/TXT/?uri=CELEX%3A52020DC0381.

© Burleigh Dodds Science Publishing Limited, 2025. All rights reserved.

EUROSTAT. (2007). The use of plant protection products in the European Union. Edition. http://ec.europa.eu/eurostat (accessed 30 June 2023).

Falk, S. P., et al. (1995). Partial control of grape powdery mildew by the mycoparasite *Ampelomyces quisqualis*. *Plant Disease*, *79*(5), 483–490.

Falk, S. P., et al. (1996). *Fusarium proliferatum* as a biocontrol agent against grape downy mildew. *Phytopathology*, *86*(10), 1010–1017.

Fedele, G., et al. (2020a). Biocontrol of *Botrytis cinerea* on grape berries as influenced by temperature and humidity. *Frontiers in Plant Science*, *11*, 1–11. https://doi.org/10.3389/fpls.2020.01232

Fedele, G., et al. (2020b). Reduction of *Botrytis cinerea* colonization of and sporulation on bunch trash. *Plant Disease*, *104*(3), 808–816.

Fedele, G., et al. (2020c). Influence of environment on the biocontrol of *Botrytis cinerea*: a systematic literature review. In A. De Cal, et al. (Eds.), *How research can stimulate the development of commercial biological control against plant diseases* (pp. 61–82). Cham: Springer. https://doi.org/10.1007/978-3-030-53238-3_5

Fedele, G., et al. (2020d). A generic model accounting for the interactions among pathogens, host plants, biocontrol agents, and the environment, with parametrization for *Botrytis cinerea* on grapevines. *Agronomy*, *10*(2), 222. https://doi.org/10.3390/agronomy10020222

Gadoury, D. M., et al. (2012). Grapevine powdery mildew (*Erysiphe necator*): a fascinating system for the study of the biology, ecology and epidemiology of an obligate biotroph. *Molecular Plant Pathology*, *13*(1), 1–16.

Gadoury, D. M., et al. (1994). Reduction of powdery mildew and other diseases by over-the-trellis applications of lime sulfur to dormant grapevines. *Plant Disease*, *78*(1), 83–87.

Gadoury, D. M., et al. (2003). Ontogenic resistance to powdery mildew in grape berries. *Phytopathology*, *93*(5), 547–555.

Gessler, C., et al. (2011). *Plasmopara viticola*: a review of knowledge on downy mildew of grapevine and effective disease management. *Phytopathologia Mediterranea*, *50*(1), 3–44.

Ghule, M. R., et al. (2018). Identification of *Fusarium* species as putative mycoparasites of *Plasmopara viticola* causing downy mildew in grapevines. *Australasian Plant Disease Notes*, *13*, 1–6.

González-Domínguez, E., et al. (2015). A mechanistic model of *Botrytis cinerea* on grapevines that includes weather, vine growth stage, and the main infection pathways. *PloS One*, *10*, 1–23. https://doi.org/10.1371/journal.pone.0140444

González-Domínguez, E., et al. (2019). A network meta-analysis provides new insight into fungicide scheduling for the control of *Botrytis cinerea* in vineyards. *Pest Management Science*, *75*(2), 324–332.

Gouadec, D. & Blouin, J. (2007). *Les Parasites de la Vigne: Stratégies de Protection Raisonnée Stratégies de Protection Raisonnée*. Paris, France: Dunod.

Gramaje, D., et al. (2018). Managing grapevine trunk diseases with respect to etiology and epidemiology: current strategies and future prospects. *Plant Disease*, *102*, 12–39. https://doi.org/10.1094/PDIS-04-17-0512-FE

Hamdi, R. F. & Alzawi, K. S. (2023). Using fungi and bacteria as biological control agents of fungal plant diseases an article. Journal of University of Anbar for Pure Science, *17*(1), 50–61.

Hao, Z., et al. (2017). Biocontrol of grapevine aerial and root pathogens by *Paenibacillus* sp. strain B2 and paenimyxin in vitro and in planta. *Biological Control*, *109*, 42–50.

© Burleigh Dodds Science Publishing Limited, 2025. All rights reserved.

He, D. C., et al. (2021). Biological control of plant diseases: An evolutionary and eco-economic consideration. *Pathogens, 10*(10), 1311.

Higgins, J. P. T. & Thompson, S. G. (2002). Quantifying heterogeneity in a meta-analysis. *Statistics in Medicine, 21*, 1539–1558.

Hinckley, E. L. S., et al. (2020). A shift in sulfur-cycle manipulation from atmospheric emissions to agricultural additions. *Nature Geoscience, 13*(9), 597–604.

Hoch, H. C. & Abawi, G. S. (1979). Mycoparasitism of oospores of *Pythium ultimum* by *Fusarium merismoides*. *Mycologia, 71*(3), 621–625.

Hunter, J. E. & Schmidt, F. L. (2004). *Methods of meta-analysis: Correcting error and bias in research findings* (2nd ed.) Thousand Oaks, CA: Sage Publications Inc.

Jaber, L. R. (2015). Grapevine leaf tissue colonization by the fungal entomopathogen *Beauveria bassiana* sl and its effect against downy mildew. *BioControl, 60*, 103–112.

Jackson, D., et al. (2012). Quantifying the impact of between-study heterogeneity in multivariate meta-analyses. *Statistics in Medicine, 31*, 3805–3820.

Jelenić, J., et al. (2021). Endophytic *Fusarium* spp. affect yield, but can they be a solution for *Botrytis* control in vineyard production? *Fungal Ecology, 54*, 101114.

Ji, T., et al. (2023). Temperature-dependent growth and spore germination of fungi causing grapevine trunk diseases: Quantitative analysis of literature data. *Plant Disease, 107*(5), 1386–1398.

John, S., et al. (2005). Protection of grapevine pruning wounds from infection by *Eutypa lata* using *Trichoderma harzianum* and *Fusarium lateritium*. *Australasian Plant Pathology, 34*, 569–575.

Keele, S. (2007). Guidelines for performing systematic literature reviews in software engineering. Technical Report, EBSE Technical Report EBSE-2007-01, 1-57.

Kennelly, M. M., et al. (2005). Seasonal development of ontogenic resistance to downy mildew in grape berries and rachises. *Phytopathology, 95*(12), 1445–1452.

Khan, N. I., et al. (1993). Parasitism of oospores of *Pythium* spp. by strains of Actinoplanes spp. *Canadian Journal of Microbiology, 39*(10), 964–972.

Kiss, L., et al. (2004). Biology and biocontrol potential of *Ampelomyces mycoparasites*, natural antagonists of powdery mildew fungi. *Biocontrol Science and Technology, 14*(7), 635–651.

Kongue Tatong, M. D., et al. (2014). Banchromene and other secondary metabolites from the endophytic fungus Fusarium sp. obtained from Piper guineense inhibit the motility of phytopathogenic *Plasmopara viticola* zoospores. *Tetrahedron Letters, 55*(30), 4057–4061.

Konstantinidou-Doltsinis, S., et al. (2007). Control of powdery mildew of grape in Greece using Sporodex® L and Milsana®. *Journal of Plant Diseases and Protection, 114*, 256–262.

Koricheva, J. & Gurevitch, J. (2014). Uses and misuses of meta-analysis in plant ecology. *Journal of Ecology, 102*(4), 828–844.

Kortekamp, A. (1997). *Epicoccum nigrum* LINK: a biological control agent of *Plasmopara viticola* (BERK. et CURT.). *Vitis, 36*(4), 215–216.

Kotze, C., et al. (2011). Evaluation of biocontrol agents for grapevine pruning wound protection against trunk pathogen infection. *Phytopathologia Mediterranea, 50*, S247–S263.

Kumar, A. & Droby, S. (2021). *Microbial management of plant stresses: current trends, application and challenges* (First eds. A. Kumar and S. Droby). Cambridge: Woodhead Publishing.

© Burleigh Dodds Science Publishing Limited, 2025. All rights reserved.

Lahlali, R., et al. (2022). Biological control of plant pathogens: A global perspective. *Microorganisms*, *10*(3), 596.

Lappa, I. K., et al. (2018). Control of *Aspergillus carbonarius* in grape berries by *Lactobacillus plantarum*: a phenotypic and gene transcription study. *International Journal of Food Microbiology*, *275*, 56–65.

Legler, S. E., et al. (2011). Effect of different plant protection products on the sexual stage of grapevine powdery mildew. In *Proceedings of the IOBC/WPRS European meeting of the working group" Integrated Protection and Production in Viticulture"* (pp. 2–2). N/A.

Legler, S. E., et al. (2012). A nonlinear model for temperature-dependent development of *Erysiphe necator* chasmothecia on grapevine leaves. *Plant Pathology*, *61*(1), 96–105.

Legler, S. E., et al. (2016). Sporulation rate in culture and mycoparasitic activity, but not mycohost specificity, are the key factors for selecting *Ampelomyces* strains for biocontrol of grapevine powdery mildew (Erysiphe necator). *European Journal of Plant Pathology*, *144*, 723–736.

Leifert, C., et al. (1995). Antibiotic production and biocontrol activity by *Bacillus subtilis* CL27 and *Bacillus pumilus* CL45. *Journal of Applied Microbiology*, *78*(2), 97–108.

Lemos Junior, W. J. F., et al. (2016). Biocontrol ability and action mechanism of Starmerella bacillaris (synonym Candida zemplinina) isolated from wine musts against gray mold disease agent Botrytis cinerea on grape and their effects on alcoholic fermentation. *Frontiers in Microbiology*, *7*, 1249.

Lo Piccolo, S., et al. (2015). Identification of *Acremonium* isolates from grapevines and evaluation of their antagonism towards Plasmopara viticola. *Annals of Microbiology*, *65*, 2393–2403.

Loevinsohn, M. (1987). Insecticide use and increased mortality in rural Central Luzon, Philippines. *The Lancet*, *329*(8546), 1359–1362.

Maachia, B. S., et al. (2015). Biological control of the grapevine diseases 'grey mold' and 'powdery mildew' by *Bacillus* B27 and B29 strains.

Madden, L. V., and Paul, P. A. (2011). Meta-analysis for evidence synthesis in plant pathology: An overview. *Phytopathology*, 101(1), 16–30.

Magarey, P. & Moyer, M. (2010, July). Towards establishing slow input regimes in Australian viticulture 3: Use of "epi-season" and" lag phase control" in applying epidemiological knowledge of grapevine powdery mildew, to reduce the number of sprays and inoculum reservoirs for long-term control. In *Proceedings of the 6th International Workshop on Grapevine Downy and Powdery Mildew, Bordeaux, France* (pp. 4–9).

Magnin-Robert, M., et al. (2013). Differential induction of grapevine resistance and defense reactions against *Botrytis cinerea* by bacterial mixtures in vineyards. *BioControl*, *58*, 117–131.

Markellou, E., et al. (2022). Evaluation of plant protection efficacy in field conditions and side effects of Lysobacter capsici AZ78, a biocontrol agent of *Plasmopara viticola*. *Biocontrol Science and Technology*, *32*(8), 930–951.

Marrone, P. G. (2007). Barriers to adoption of biological control agents and biological pesticides. *CABI Reviews*, *2007*, 12-pp.

Marsico, A. D., et al. (2021). Native vineyard non-*Saccharomyces* yeasts used for biological control of *Botrytis cinerea* in stored table grape. *Microorganisms*, *9*(2), 457.

© Burleigh Dodds Science Publishing Limited, 2025. All rights reserved.

Mohamed, N., et al. (2007). Defense responses in grapevine leaves against *Botrytis cinerea* induced by application of a *Pythium oligandrum* strain or its elicitin, oligandrin, to roots. *Phytopathology, 97*(5), 611–620.

Molitor, D. & Berkelmann-Loehnertz, B. (2011). Simulating the susceptibility of clusters to grape black rot infections depending on their phenological development. *Crop Protection, 30*(12), 1649–1654.

Mondello, V., et al. (2018). Grapevine trunk diseases: a review of fifteen years of trials for their control with chemicals and biocontrol agents. *Plant Disease, 102*(7), 1189–1217.

Musetti, R., et al. (2006). Inhibition of sporulation and ultrastructural alterations of grapevine downy mildew by the endophytic fungus *Alternaria alternata*. *Phytopathology, 96*(7), 689–698.

Mutawila, C., et al. (2015). Development of benzimidazole resistant *Trichoderma* strains for the integration of chemical and biocontrol methods of grapevine pruning wound protection. *BioControl, 60*, 387–399.

Nally, M. C., et al. (2015). Antifungal modes of action of *Saccharomyces* and other biocontrol yeasts against fungi isolated from sour and grey rots. *International Journal of Food Microbiology, 204*, 91–100.

Ngugi, H. K., et al. (2011). Meta-analysis to determine the effects of plant disease management measures: review and case studies on soybean and apple. *Phytopathology, 101*(1), 31–41.

Niem, J. M., et al. (2023). Biocontrol potential of an endophytic *Pseudomonas poae* strain against the grapevine trunk disease pathogen *Neofusicoccum luteum* and its mechanism of action. *Plants, 12*(11), 2132.

Niem, J. M., et al. (2020). Diversity profiling of grapevine microbial endosphere and antagonistic potential of endophytic *Pseudomonas* against grapevine trunk diseases. *Frontiers in Microbiology, 11*, 477.

Noordzij, M., et al. (2009). Systematic reviews and meta-analyses: when they are useful and when to be careful. *Kidney international, 76*(11), 1130–1136.

Okoli, C., and Schabram, K. (2010). A guide to conducting a systematic literature review of information systems research. *Papers Inf. Syst.* 10, 1–46. doi: 10.2139/ssrn.1954824

Parnell, J. J., et al. (2016). From the lab to the farm: an industrial perspective of plant beneficial microorganisms. *Frontiers in Plant Science, 7*, 1110.

Pascale, A., et al. (2017). *Trichoderma* and its secondary metabolites improve yield and quality of grapes. *Crop protection, 92*, 176–181.

Pedrotti, C., et al. (2020). Essential oil as sustainable alternative for diseases management of grapes in postharvest and in vineyard and its influence on wine. *Journal of Environmental Science and Health, Part B, 56*(1), 73–81.

Perrone, G., et al. (2004). PCR assay for identification of *Aspergillus carbonarius* and *Aspergillus japonicus. Molecular Diversity and PCR-detection of Toxigenic Fusarium Species and Ochratoxigenic Fungi: Under the aegis of COST Action 835 'Agriculturally Important Toxigenic Fungi 1998–2003', EU project (QLK1-CT-1998--01380) and the ISPP 'Fusarium Committee'*, 641–649.

Pertot, I., et al. (2016). *Trichoderma atroviride* SC1 prevents *Phaeomoniella chlamydospora* and *Phaeoacremonium aleophilum* infection of grapevine plants during the grafting process in nurseries. *BioControl, 61*, 257–267.

Pertot, I., et al. (2017). A critical review of plant protection tools for reducing pesticide use on grapevine and new perspectives for the implementation of IPM in viticulture. *Crop Protection, 97*, 70–84.

© Burleigh Dodds Science Publishing Limited, 2025. All rights reserved.

Pertot, I., et al. (2017). Combining biocontrol agents with different mechanisms of action in a strategy to control *Botrytis cinerea* on grapevine. *Crop Protection*, *97*, 85–93.

Rantsiou, K., et al. (2020). Impact of chemical and alternative fungicides applied to grapevine cv Nebbiolo on microbial ecology and chemical-physical grape characteristics at harvest. *Frontiers in Plant Science*, *11*, 700.

Raut, V. M., et al. (2018). Evaluation of Milastin-K (*Bacillus subtilis* KTSB 1015 1.5% AS) as a potential biocontrol agent for management of downy and powdery mildew in grape vines. *Pestology*, *42*(10), 15–24.

Raymaekers, K., et al. (2020). Screening for novel biocontrol agents applicable in plant disease management–a review. *Biological Control*, *144*, 104240.

Reis, P., et al. (2021). Combining an HA+ Cu (II) site-targeted copper-based product with a pruning wound protection program to prevent infection with *Lasiodiplodia* spp. in grapevine. *Plants*, *10*(11), 2376.

Rezgui, A., et al. (2016). Endophytic bacteria with antagonistic traits inhabit the wood tissues of grapevines from Tunisian vineyards. *Biological Control*, *99*, 28–37.

Ritika, B. & Utpal, D. (2014). An overview of fungal and bacterial biopesticides to control plant pathogens/diseases. *African Journal of Microbiology Research*, *8*(17), 1749–1762.

Roca, P. (2022, April). State of the world vine and wine sector. OIV *Press Conference*. https://doi.org/org/10.1787/9789264244047-44-en

Roca-Couso, R., et al. (2021). Mechanisms of action of microbial biocontrol agents against *Botrytis cinerea*. *Journal of Fungi*, *7*(12), 1045.

Ronseaux, S., et al. (2013). Interaction of *Ulocladium atrum*, a potential biological control agent, with Botrytis cinerea and grapevine plantlets. *Agronomy*, *3*(4), 632–647.

Rosace, M. C., et al. (2023). Susceptibility of pruning wounds to grapevine trunk diseases: A quantitative analysis of literature data. *Frontiers in Plant Science*, *14*, 1063932.

Rossi, V., et al. (2010). Dynamics of ascospore maturation and discharge in Erysiphe necator, the causal agent of grape powdery mildew. *Phytopathology*, *100*, 1321–1329.

Rossi, V., et al. (2012). Helping farmers face the increasing complexity of decision-making for crop protection. *Phytopathologia Mediterranea*, *51*, 457–479.

Sawant, I. S., et al. (2017). Enhanced biological control of powdery mildew in vineyards by integrating a strain of Trichoderma afroharzianum with sulphur. *Biological Control*, *114*, 133–143.

Scherm, H., et al. (2014). Meta-analysis and other approaches for synthesizing structured and unstructured data in plant pathology. *Annual Review of Phytopathology*, 52(1), 453–476.

Silva-Valderrama, I., et al. (2021). Biocontrol potential of grapevine endophytic and rhizospheric fungi against trunk pathogens. *Frontiers in microbiology*, *11*, 614620.

Singh, A., et al. (2020). Biological control of plant diseases: opportunities and limitations. In: Varma, A., et al. (Eds), *Plant Microbiome Paradigm*, 121–146. https://doi.org/10.1007/978-3-030-50395-6_7

Sipiczki, M. (2006). Metschnikowia strains isolated from botrytized grapes antagonize fungal and bacterial growth by iron depletion. *Applied and Environmental Microbiology*, *72*(10), 6716–6724.

Sneh, B., et al. (1977). Parasitism of oospores of *Phytophthora megasperma* var. *sojae*, *P. cactorum*, *Pythium* sp., and *Aphanomyces euteiches* in soil by oomycetes,

© Burleigh Dodds Science Publishing Limited, 2025. All rights reserved.

chytridiomycetes, hyphomycetes, actinomycetes, and bacteria. *Phytopathology*, *67*(5), 622–628.

Sutherland, E. D. & Lockwood, J. L. (1984). Hyperparasitism of oospores of some peronosporales by *Actinoplanes missouriensis* and *Humicola fuscoatra* and other Actinomycetes and fungi. *Canadian Journal of Plant Pathology*, *6*(2), 139–145.

Thiessen, L. D., et al. (2018). Interruption and reduction of *Erysiphe necator* chasmothecia development utilizing fungicidal oil. *Plant Health Progress*, *19*(2), 153–155.

Todaro T. & Miles T. (2018). Early season vineyard disease management. https://www.canr.msu.edu/news/early_season_vineyard_disease_management (accessed 20 July 2023).

Toral, L., et al. (2018). Antifungal activity of lipopeptides from *Bacillus* XT1 CECT 8661 against *Botrytis cinerea*. *Frontiers in Microbiology*, *9*, 1315.

Tracy, E. F. (2014). The promise of biological control for sustainable agriculture: a stakeholder-based analysis. *Journal of Science Policy & Governance*, *5*(1), 1–13.

Tran, C., et al. (2022). Antimicrobial *Bacillus*: metabolites and their mode of action. *Antibiotics*, *11*(1), 88.

Trotel-Aziz, P., et al. (2008). Characterization of new bacterial biocontrol agents *Acinetobacter*, *Bacillus*, *Pantoea* and *Pseudomonas* spp. mediating grapevine resistance against *Botrytis cinerea*. *Environmental and Experimental Botany*, *64*(1), 21–32.

Van Helden, M. (2008). Protection intégrée: La protection intégrée vis-à-vis des ravageurs de la vigne. *Ravageurs de la vigne*. Féret, France, 321–335.

Vanderplank, J. E. (1963). *Plant diseases*. New York: Elsevier Science.

Vecchione, A., et al. (2004). Screening and identification of potential biocontrol agents against grapevine downy mildew considering an integrated control strategy of the disease. *S. Michele all'Adige, Italy*, *27*(8), 85–88.

Vecchione, A., et al. (2007). Biological control of *Plasmopara viticola*: a multisite approach. *Acta Horticulturae*, *754*, 361–366. https://doi.org/10.17660/ActaHortic.2007.754.47

Velivelli, S. L., et al. (2014). Biological control agents: from field to market, problems, and challenges. *Trends in Biotechnology*, *32*(10), 493–496.

Verger, P. J. P. and Boobis, A. R. (2013). Reevaluate pesticides for food security and safety. *Science*, *341*, 717–718. https://doi.org/10.1126/science.1241572

Viechtbauer, W. (2010). Conductingmeta-analyses in R with the metafor package. *Journal of Statistical Software*, *36*, 1–48.

Whiteman, S. A. & Stewart, A. (1998). *Suppression of Botrytis cinerea sporulation on irradiated grape leaf tissue by the antagonistic bacterium Serratia liquefaciens*. Netherlands: Springer.

Wynn, A. R. & Epton, H. A. S. (1979). Parasitism of oospores of *Phytophthora erythroseptica* in soil. *Transactions of the British Mycological Society*, *73*(2), 255–259.

Xiao, Y. & Watson, M. (2019). Guidance on conducting a systematic literature review. *Journal of Planning Education and Research*, *39*(1), 93–112.

Zhao, H., et al. (2022). Cell-free supernatant of *Bacillus velezensis* suppresses mycelial growth and reduces virulence of *Botrytis cinerea* by inducing oxidative stress. *Frontiers in Microbiology*, *13*, 980022.

© Burleigh Dodds Science Publishing Limited, 2025. All rights reserved.

Zhou, Q., et al. (2020). Application of antagonist *Bacillus amyloliquefaciens* NCPSJ7 against *Botrytis cinerea* in postharvest Red Globe grapes. *Food Science & Nutrition*, *8*(3), 1499–1508.

Zhu, C., et al. (2015). Inhibition of the growth and ochratoxin A production by Aspergillus carbonarius and *Aspergillus ochraceus* in vitro and in vivo through antagonistic yeasts. *Food Control*, *50*, 125–132.

© Burleigh Dodds Science Publishing Limited, 2025. All rights reserved.

Chapter 4

Use of biocontrol agents in fruit tree disease management

Jürgen Köhl, Wageningen University and Research, The Netherlands

1 Introduction

Biological control of plant diseases is the suppression of populations of plant pathogens by living organisms (Heimpel and Mills, 2017). Fungi, bacteria and viruses and members of the mesofauna can be involved in such natural suppression. The effect of naturally present beneficial consortia may be limited when broad-spectrum fungicides are applied that may affect these beneficial populations. Such undesired side effects of fungicides may result in the emergence of 'iatrogenic' diseases (Fokkema and de Nooij, 1981). 'Natural biological control' buffers cropping systems and lowers risks for disease outbreaks (Eilenberg et al., 2001; Heimpel and Mills, 2017). This natural biological control can be facilitated by 'conservation biological control' which stimulates the natural microbial buffering capacity or protects natural antagonists by avoiding unnecessary fungicide applications. Application of modern meta-genomic approaches allows a much better understanding of the beneficial functions of the microbiomes naturally present in and on crops (Berg, 2009). It is expected that this knowledge will lead to new management options for the use of natural biological control to ensure the suppression of pathogens. Amongst beneficial microorganisms, isolates can be selected which are highly effective against a targeted pathogen and can efficiently be multiplied on artificial media. Application of such selected and mass-produced antagonists in high densities once, or several times, during a growing season is called

http://dx.doi.org/10.19103/AS.2019.0046.16
© Burleigh Dodds Science Publishing Limited, 2019. All rights reserved.

'augmentative biological control' (Eilenberg et al., 2001; Heimpel and Mills, 2017; van Lenteren et al., 2018). Research on augmentative biological control leading to commercial products for disease control in fruit tree production is reviewed in this chapter.

Commercial biocontrol agents (BCA) for disease control are registered plant protection products, which are manufactured by specialized companies. Such products contain living microbial organisms such as fungi or bacteria. In some cases, metabolites produced by these microbial organisms are included in the product (Glare et al., 2012). Mycoviruses and bacteriophages can also be potential BCAs. The number of registered BCAs for disease control in fruit tree production in Australia, Brazil, Canada, European Union, Japan, New Zealand and the United States is limited: only fifteen antagonistic strains were registered belonging to four fungal genera and three bacterial genera (Table 1). These registered strains are targeted against a range of pathogens. Some strains are registered against several pathogens and, in some cases, registrations have been obtained for different pathogens in different regions. However, the majority of registered antagonists are targeted against fireblight (*Erwinia amylovora*), with seven strains, and pre- and post-harvest fruit rot diseases caused by various pathogens with eight strains.

Antagonists protect crops from damage by diseases via different modes of action. BCAs may induce resistance against infections by a pathogen in plant tissue without any direct antagonistic interaction with the pathogen. Two different pathways in the plant are involved in systemic acquired resistance (SAR) or induced systemic resistance (ISR). Also, microorganisms applied as biofertilizers or biostimulants and mycorrhizal fungi, not all of which are considered as biocontrol agents, can strengthen crops, leading to increased resilience against disease. Another indirect interaction with pathogens is in the competition for nutrients and space. Necrotrophic plant pathogens often need exogenous nutrients before being able to infect a host plant. Microorganisms with high saprophytic competitive ability can rapidly consume the nutrient sources essential for pathogen infection such as sugars, pollen and plant exudates on plant surfaces and the consequently outcompeted pathogens are not able to infect the host. An example of this efficient mode of action is the protection of wounds on fruits from pathogen invasion by fast colonizing yeasts (Spadaro and Droby, 2016). Antagonists may also interact directly with the pathogen by hyperparasitism or antibiosis. Hyperparasites are able to invade the pathogen mycelium and spores after enzymatic cell wall degradation by β-1,3-glucanases and chitinases, resulting in eradication of the invading pathogen cells. Production of antimicrobial secondary metabolites with toxic effects against pathogens is another mode of action leading to eradication of the targeted pathogen. Low amounts of *in situ* secreted secondary metabolites support antagonists to attack the pathogens, leading to a competitive

© Burleigh Dodds Science Publishing Limited, 2019. All rights reserved.

Table 1 Registered bacterial and fungal biological control agents for control of diseases in fruit tree crops[a]

Microbial strain	Targeted pathogens or fruit diseases	Countries where approved
Aureobasidium pullulans DSM 14940 and 14941	Fire blight (*Erwinia amylovora*) in pome fruit; brown rot in stone fruit	Europe, Canada, USA
Bacillus licheniformis Strain SB3086	Apple scab (*Venturia inaequalis*); rust (*Gymnosporangium* spp.)[b]	USA
Bacillus pumilus QST 2808	Fire blight (*Erwinia amylovora*) in pome fruit; scab (*Venturia* spp.), powdery mildew (*Podosphaera leuchotricha*)[c]	USA
Bacillus subtilis QST713	Fire blight (*Erwinia amylovora*) in pome fruit; scab (*Venturia* spp.) *Monilinia laxa*, *Monilinia fructicola*, *Podosphaera leucotricha*[d]	USA
Candida oleophila I-182	Wound diseases of pome fruits	USA
Candida oleophila O	Wound diseases of pome fruits	EU
Metschnikowia fructicola NRRL Y027328	Wound diseases of pome fruits	EU (pending)
Pantoea agglomerans C9-1	Fire blight (*Erwinia amylovora*) in pome fruit	CA, USA
Pantoea agglomerans E325	Fire blight (*Erwinia amylovora*) in pome fruit	CA
Pantoea agglomerans p10c	Fire blight (*Erwinia amylovora*) in pome fruit	NZ
Pseudomonas fluorescens A506	Fire blight (*Erwinia amylovora*) in pome fruit	USA
Pseudomonas rhodesiae HAI-0804	*Pseudomonas syringae* in plum[e]	Japan
Pseudomonas syringae ESC 10	Post-harvest fungal diseases of fruits of apples, pears and cherries[f]	CA, USA
Pseudomonas syringae ESC 11	Various moulds including blue mould (*P. expansum*), grey mould (*B. cinerea*) and Mucor rot on fruits of apple, pear and cherries[g]	USA
Saccharomyces cerevisiae LAS02	Fruit rots caused by *Monilinia fructigena*, *Monilinia laxa*, *Monilinia fructicola*, *Botrytis* spp., *Botrytis cinerea*, *Alternaria* spp. *Neofabraea alba* and *Penicillium* spp.[h]	EU

[a] For Australia, Brazil, Canada, European Union, Japan, New Zealand and the United States (from: van Lenteren et al., 2018).
[b] https://www3.epa.gov/pesticides/chem_search/.../decision_PC-006492_1-Feb-01.pdf
[c] https://www3.epa.gov/pesticides/chem_search/reg_actions/registration/decision_PC-006485_16-Nov-04.pdf
[d] https://www3.epa.gov/pesticides/chem_search/reg_actions/registration/decision_PC-006479_9-Aug-06.pdf
[e] http://www.abim.ch/fileadmin/abim/documents/presentations2016/3_Tetsuo_Wada_ABIM_2016.pdf
[f] http://sitem.herts.ac.uk/aeru/bpdb/Reports/1998.htm
[g] http://sitem.herts.ac.uk/aeru/bpdb/Reports/1999.htm
[h] http://sitem.herts.ac.uk/aeru/bpdb/Reports/2954.htm

© Burleigh Dodds Science Publishing Limited, 2019. All rights reserved.

advantage of the antagonist. Biocontrol agents have also been selected which secrete efficient secondary metabolites into the growth media during mass production and secondary metabolites are applied together with living cells of the antagonists in the commercial biocontrol product.

For the development of new BCA, the choice of the mode of action depends on the targeted disease, for example nutrient competition will not affect biotrophic pathogens, as well as the targeted stage of the pathogen life cycle. However, many more criteria besides the preferred modes of action have to be considered during the screening of new antagonists. Such criteria include ecological characteristics such as cold tolerance and UV resistance, and economically important characteristics such as high spore production potential to allow mass production. Also, potential risks have to be considered to exclude strains with potential pathogenicity to humans, animals or plants. Screening of antagonists for commercial use is thus a complex process that focusses not only on the efficacy of antagonists and their mode of action, but also on various other relevant criteria in a stepwise approach (Köhl et al., 2011).

2 Case study: strategies for biocontrol of apple scab - *Venturia inaequalis*

Apple scab, caused by *Venturia inaequalis*, is the most important disease worldwide affecting apple production. The life cycle of the pathogen consists of two distinct parts (MacHardy et al., 2001). During winter, the pathogen survives on apple leaf litter with a saprophytic lifestyle. Interestingly, the pathogen colonizes necrotic apple leaves but is not able to colonize other necrotic plant tissues. Survival of conidia on twigs is a second, less important, way of overwintering. The main disease source inciting new epidemics in spring comes from the ascospores produced on overwintering apple leaves. Ascospores infect the young tissues of developing leaves and young fruits. Thereafter, a polycyclic epidemic develops with many infection-sporulation cycles through conidia. During the summer part of the life cycle, the pathogen grows with a semi-biotrophic lifestyle on epidermis cells under the cuticles of leaves and fruits. The particular life cycle of *V. inaequalis* determines the options and limitations for biological control of apple scab. During the winter and summer parts of the life cycle there are differences in the lifestyle of the pathogen (saprophytic versus semi-biotrophic), niche (leaf litter on the ground versus canopy) and colonization of host tissue (internal and external colonization of dead tissue versus subcuticular colonization of living tissue), and as such, biocontrol requires different types of antagonists during these two seasons.

During the winter season, antagonists with superior saprophytic competitive colonization ability may reduce the survival of *V. inaequalis* mycelium and the formation of ascospores in fruiting bodies. Since no alternative necrotic

© Burleigh Dodds Science Publishing Limited, 2019. All rights reserved.

plant tissues can be colonized by *V. inaequalis*, this application can be very well targeted at leaf litter on the orchard floor to reach potential pathogen sources. Possible ecological restrictions of this biocontrol strategy are the low temperatures during autumn and winter and the resistance of *V. inaequalis* fruiting bodies to antagonism. An epidemiological limitation of this strategy is that sanitation measures, including the removal of leaf litter, shredding of leaf litter or application of nutrients to stimulate naturally occurring microbial decomposers, are an important component in apple scab management which can delay the build-up of epidemics in spring but are not sufficient as stand-alone measures since under practical conditions, the pathogen load cannot be completely eliminated (Sutton et al., 2000; Carisse and Dewdney, 2002; Gomez et al., 2007).

The second option for biocontrol is the application of antagonists during the summer epidemic. Such antagonists may interfere with *V. inaequalis* before and during infection on the host tissue, or later in the cycle during colonization of the sub-epidermal space and subsequent pathogen sporulation. Ungerminated ascospores and conidia of *V. inaequalis* are present on the host surface during periods of low humidity, with infection processes depending on leaf wetness and temperature. Antagonists applied to prevent infections must be able to survive on host tissue surfaces during the period from application until infection. Under low humidity conditions, metabolites present in the antagonist formulation may interact with *V. aequalis*; however, during leaf wetness periods, antagonists must become active and be able to interact antagonistically with *V. aequalis*. Since *V. aequalis* spores can germinate and penetrate the cuticle within a few hours, activation of the antagonist's metabolism and interaction with the host must be sufficiently fast to achieve successful biocontrol. In cases where interaction with *V. inaequalis* occurs during colonization and sporulation in the subcuticular space, antagonists must be able to grow into this space without becoming pathogenic or causing skin disorders. A third opportunity for biocontrol during the summer epidemic may be the use of resistance-inducing antagonists (Ortega et al., 1998; Percival et al., 2009).

Biocontrol of apple scab has been an area of investigation for the past several decades (Carisse, 2000). Heye and Andrews (1983) screened fungal antagonists for their potential to reduce pseudothecium formation of *V. inaequalis* in naturally infected senesced apple leaves. The selected basiodiomycete, *Athelia bombacina,* inhibited pseudothecium formation efficiently also under orchard conditions. The potential of *A. bombacina* was confirmed by Miedtke and Kennel (1990), although a major bottleneck in this case is the required high amount of inoculum of the antagonist for orchard applications, given the difficulties in mass production of the basidiomycete (Carisse and Dewdney, 2002). Autumn application of *Microsphaeropsis ochracea*, a fungal antagonist with high potential in mass production, to the orchard floor was found to reduce

© Burleigh Dodds Science Publishing Limited, 2019. All rights reserved.

the level of ascospores of the pathogen during the following spring by 70–80% (Carisse et al., 2000), allowing a reduction in the number of fungicide sprays employed during the cropping season.

The use of antagonists during the summer epidemics was first investigated by Andrews et al. (1983) who compared 50 isolates belonging to different species of bacteria and fungi. *Chaetomium globosum* had the strongest effect on *V. inaequalis* in tests under controlled conditions. However, the antagonist was not able to colonize leaves under orchard conditions and the antibiotic produced by the antagonist was not stable in the environment (Boudreau and Andrews, 1987). Burr et al. (1996) selected antagonists affecting the conidial germination and mycelial growth of *V. inaequalis*. Amongst the almost 1000 isolates tested, an isolate of *Pseudomonas syringae* was found to be superior and reduced infection by the pathogen on seedlings. Epiphytic yeasts were selected by Fiss et al. (2000) in a screening programme but no reproducible control effects were achieved under orchard conditions.

In a screening programme using bioassays with apple seedlings artificially inoculated with conidia of *V. inaequalis*, a new antagonist was discovered (Köhl, 2009; Köhl et al., 2009). *Cladosporium cladosporioides* H39, isolated from an apple leaf collected in a Dutch orchard, reduced conidial production by *V. inaequalis* when applied one to three days after pathogen inoculation. A pilot formulation of conidia of *C. cladosporioides* H39 was used in eight field trials in 2012 and 2013 in the main apple growing areas in Germany, Hungary and Poland (Köhl et al., 2015). These trials were carried out in conventional cropping systems as well as in organic systems and focussed on apple scab control during the primary or summer season. Control efficacies ranged between 42% and 98% on leaf scab and from 41% to 94% on fruit scab. In some cases the efficacies achieved were no different from those obtained by application of synthetic fungicides or copper-based fungicides, especially when applied during the summer season (Fig. 1). Interestingly, the antagonist was also effective when applied at 300–2000 degree hours after forecasted infection events, suggesting that *V. aequalis* and *C. cladosporioides* H39 can also interact after spore germination during the subcuticular colonization.

Biocontrol products for scab control are currently commercially available only in the United States, where three *Bacillus* spp. strains are registered with claimed effects on *Venturia* spp. (Table 1). Registration of new biocontrol products for scab control will have several advantages for scab management. Substitution of applications of synthetic fungicides will lower environmental risks and reduce the number and concentration of residues of synthetic fungicides on fruits. Furthermore, the use of BCA will allow a better management of strains of *V. inaequalis* with resistance to fungicides in orchard conditions (Beresford et al., 2013). In the scenario in which the number of registered chemical fungicides available for scab control decreases, and given the preference

© Burleigh Dodds Science Publishing Limited, 2019. All rights reserved.

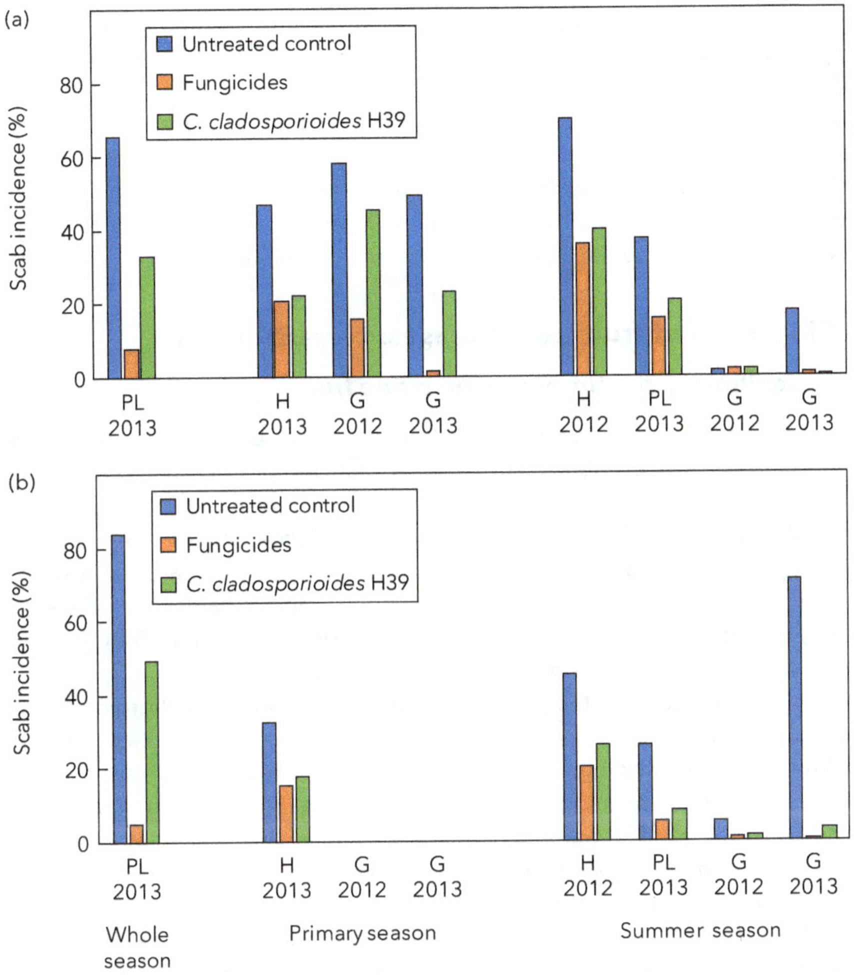

Figure 1 Effect of treatments with *Cladosporium cladosporioides* H39 in comparison to fungicides applied at regular intervals on (a) scab on leaves and (b) scab on fruit. Field trials in Poland (PL), Hungary (H) and Germany (G). Source: Adapted from Köhl et al. (2015).

for low-risk plant protection, products such as BCA will become an essential component of scab management in commercial apple production.

Scab management must fit into the management strategies applied to control other diseases and pests. The use of fungicides for the control of powdery mildew, various fruit rots or against fungi causing superficial disorders on fruits must be compatible with the use of BCA applied for scab control. There is no need to achieve full compatibility however, allowing applications with tank mixes. Established antagonist populations may be reduced by chemical

© Burleigh Dodds Science Publishing Limited, 2019. All rights reserved.

fungicides applied alternately with BCA application, although antagonist populations can be re-established by subsequent application after a few days when the fungicides have lost their activity or are washed off. Alternations in the use of chemical and biocontrol products will thus offer many options for the application of biocontrol within integrated pest management systems. For the development of such alternating application schedules, knowledge of fungicide persistence and the population dynamics of the antagonist under fluctuating environmental conditions in the orchard are essential.

3 Biocontrol of fruit tree diseases: current status

3.1 Monilinia *spp. – Brown rot on stone fruit*

Monilinia fructicola, *M. laxa* and *M. fructigena* cause brown rot, the most important disease on stone fruit. The epidemiology and management of brown rot caused by *M. laxa* have recently been reviewed by Rungjindamai et al. (2014a). Infected mummies are the main substrates for overwintering and disease sources for conidium production in spring. Flowers and developing fruits can be infected by *M. fructicola* and *M. laxa*, and at harvest, the fruits are highly susceptible and infections through wounds are common. Fruit damage by brown rot occurs during the season in the orchard and during storage, where latent infections in the pre-harvest phase and wound infections caused at harvest lead to brown rot, which may further spread through contact. Infection of fruit via wounds both pre- and post-harvest can lead to the rapid rotting of the fruit. Brown rot epidemiology favours three strategies for biocontrol: microbial reduction of pathogen build-up in overwintering mummies, pre-harvest prevention of flower and fruit infection in the orchard and post-harvest prevention of wound infection before storage.

Biocontrol on harvested fruit before storage has been studied by many research groups and a broad range of bacteria and yeasts have been tested for this application. Screening of new antagonists is straightforward in this case because bioassays with artificially wounded fruit mimic practical conditions with a well-defined time window for application during a known infection period and under well-defined environmental conditions. Assessment of symptoms is also fast and directly linked to losses caused by infections. The entire fruit surface can easily be reached by antagonists via dipping and the surface area for treatment is small compared to spray applications in orchards. Progress in the use of antagonists in post-harvest control has recently been reviewed by Rungjindamai et al. (2014a). Several isolates of *Bacillus subtilis* and *B. amyloliquefaciens* have successfully been used for post-harvest brown rot control (Pusey et al., 1988; Zhou et al., 2008; Casals et al., 2010). Lipopeptides such as fengycin, produced by strains of *Bacillus* spp., are inhibitory to fungal mycelial growth and protect treated fruit wounds from invasion by *Monilinia*

© Burleigh Dodds Science Publishing Limited, 2019. All rights reserved.

spp. *Bacillus subtilis* QST713 has been registered for brown rot control in the United States (Table 1). Yeasts and yeast-like fungi can also protect stone fruit from infections by *Monilinia* spp. after harvest. *Aureobasidium*, *Cryptococcus*, *Rhodotorula* and *Sporidiobolus* are common genera colonizing fruit surfaces which can quickly colonize wound tissues and may prevent pathogen infections by nutrient competition in the wounds (Janisiewicz et al., 2010). *Aureobasidium pullulans* and *Rhodotorula phylloplana* showed excellent potential for biocontrol of brown rot caused by *M. fructicola* in plums (Janisiewicz et al., 2014). Isolates of the yeasts *Debaryomyces hansenii* and *Wickerhamomyces anomalus* have a high capacity to colonize wound tissues of peach and plum fruits and protected these fruits from infection in *in vivo* assays after artificial pathogen inoculation with *M. fructigena* and *M. fructicola* (Grzegorczyk et al., 2017). In similar studies, isolates of *Pichia membranaefaciens* and *Kloeckera apiculata* were effective against infections of plums by *M. fructicola* (Zhang et al., 2017). A product containing strains of the yeast-like fungus *A. pullulans* has been registered for brown rot control in stone fruit in the European Union, Canada and the United States.

Another strategy for biocontrol of brown rot is the application of antagonists in the orchard to prevent infections of twigs, blossoms and fruits, including latent fruit infections which can lead to damage after harvest. Melgarejo et al. (1985) assessed the fungal flora of peach twigs and flowers and amongst the 19 most abundant species, isolates of *Aspergillus flavus*, *Epicoccum nigrum*, *Penicillium chrysogenum*, *P. frequentans* and *P. purpurogenum* showed *in vitro* inhibition of *M. laxa*. Isolates of *A. flavus*, *E. nigrum*, *P. frequentans* and *P. purpurogenum* were applied to peach twigs under orchard conditions (Melgarejo et al., 1986). *Aspergillus flavus*, *E. nigrum* and *P. frequentans* inhibited twig infection by *M. laxa* in the spring when applied before inoculation with the pathogen. *Penicillium frequentans* was also effective when applied after pathogen inoculation. In similar trials in the autumn, only applications of *E. nigrum* were effective. When applied at bloom and pre-harvest, *E. nigrum* moderately reduced brown rot after harvest in five out of the seven fields (Larena et al., 2005). The use of the promising antagonist *P. frequentans* Pf909 has been further investigated, including detailed studies of field application, mode of action, mass production, formulation population dynamics in the orchard and safety (for further information see Guijarro et al., 2017).

Bacillus amyloliquefaciens CPA-8 is another efficient antagonist of *M. laxa* and *M. fructicola*. Different formulations have been tested in field trials with peaches (Gotor-Vila et al., 2017a). In orchards with high pathogen pressure, four applications of the antagonist in orchards at days 30, 14, 7 and 3 before harvest had no effect on disease incidence at harvest or in the post-harvest assessment. In such cases, synthetic fungicide only managed to achieve control levels of approximately 50%. However, the orchards used for trials with such high

© Burleigh Dodds Science Publishing Limited, 2019. All rights reserved.

pathogen pressure are not representative of commercial orchards. Applications of formulated *B. amyloliquefaciens* CPA-8 in an orchard with a moderate disease pressure significantly reduced brown rot at harvest by 55–64% and in the post-harvest assessment by 50%. Protocols for mass production and formulation of *B. amyloliquefaciens* CPA-8 have been developed and strain-specific markers are available for monitoring antagonist populations after field releases (Gotor-Vila et al., 2016, 2017b).

Rungjindamai et al. (2014b) investigated the effect of antagonists on sporulation of *M. laxa* in pathogen-infected mummies, which are a significant inoculum source for blossom and fruit infections in spring. A single application of *Aureobasidium pullulans* Y126 in winter reduced *M. laxa* sporulation as much as the fungicide fenbuconazole. Antagonist applications in early spring were slightly more effective, but the best results were obtained with a combination of winter and spring applications. *Aureobasidium pulluans* Y126 reduced the number of *M. laxa* conidia per plum by more than 90% from 9×10^5 to 5×10^4 conidia per mummified plum. Strains *Bacillus* sp. B91 and *B. subtilis* QST 713 (formulated as Serenade) showed some antagonistic activity. Such a biocontrol strategy during the dormant season can be an efficient component of integrated disease management with the advantage of long interaction periods between pathogen and antagonists from autumn until early spring. Screening for antagonists with high saprophytic competitive ability during colonization of necrotic host tissues is a generally neglected but promising strategy for biocontrol (Köhl and Fokkema, 1998). In mummified fruit, highly competitive saprophytic antagonists may be able to outcompete a whole range of fruit pathogens which overwinter in mummies. Besides *M. laxa*, other important fruit rot pathogens overwintering on mummified fruits, such as *Neofabraea alba* and *Cadophora luteo-olivacea* (Köhl et al., 2018), may also be controlled by such BCA.

3.2 Fruit rots caused by various pathogens

A range of fungal pathogens can cause fruit rot, leading to damage in the orchard or after harvest during storage. For several pathogens such as *Botrytis cinerea* (grey mould), *Penicillium expansum* (blue mould) and *Monilinia fructigena* (brown rot), wounds caused by insects and birds as well as by physical damage before or during harvest are important entry points (Snowdon, 1990). These pathogens typically result in a rapid decay of fruit in the pre- and post-harvest stages. Applications of BCA shortly before or after harvest are effective measures to reduce losses by wound pathogens because the timing of applications and the targeted parts of the crops are clearly defined and the environmental conditions after harvest are controlled and thus known. Antagonists can be screened for such applications in bioassays with artificially

© Burleigh Dodds Science Publishing Limited, 2019. All rights reserved.

wounded fruits and incubated under commercial storage conditions. Results of bioassays can thus be directly translated to commercial application conditions.

Since the successful use of *Bacillus subtilis* for brown rot control in stone fruits in the early 1980s by Pusey and Wilson (1984), much research has been directed at selecting new antagonists for fruit rot control in fruit tree crops. In a comprehensive review of biocontrol of post-harvest diseases in fruits by microbial antagonists, Sharma et al. (2009) listed 26 antagonist species which were successfully tested against the main fruit diseases in stone and pome fruits in pre- and post-harvest applications (Table 2). The majority of the antagonist species are yeasts or yeast-like fungi or bacteria. For most of the tested isolates, competition for nutrients plays a major role, although other modes of action are also involved (Piombo et al., 2018). Sharma et al. (2009) also reviewed options for enhancing the efficacy of antagonists against post-harvest diseases through adapting the physical environment during storage using antagonist mixtures or by adding low amounts of fungicides, salts such as calcium bicarbonate or sodium carbonate, specific nutrients or plant extracts. Antagonist treatments have also been successfully combined with physical treatments such as curing or heat treatment. Progress in the research of biocontrol of fruit rot diseases resulted in the development of commercial BCAs based on the antagonists *Aureobasidium pullulans* DSM 14940 and 14941, *Candida oleophila* I-182, *C. oleophila* O, *Metschnikowia fructicola* NRRL Y027328, *Pseudomonas syringae* ESC 10, *P. syringae* ESC 11 and *Saccharomyces cerevisiae* LAS02 (Table 1). Droby et al. (2016) observed reduced and inconsistent performance of biocontrol products against post-harvest fruit pathogens because commodities vary in genetic background, physiology, post-harvest handling and susceptibility to specific pathogens. This variability makes it difficult to protect commodities with one single antagonist strain. A more promising option may be the development of synthetic microbial communities to develop more consistent and robust biocontrol solutions (Droby et al., 2016; Droby and Wisniewski, 2018).

In addition to wound infecting pathogens, another important group of pathogens infects blossoms or developing fruits during the growing season and remains quiescent without causing symptoms during the growing season and during early post-harvest storage period. Typically, symptoms of disease occur after several months in cold storage within a controlled atmosphere. Common pathogens causing such late post-harvest rot are *Neofabraea alba* (lenticel rot disease), *Neofabraea perennans* (bull's eye rot), *Cadophora malorum*, *Cadophora luteo-olivacea*, *Neonectria ditissima* (blossom-end rot), *Colletotrichum acutatum* species complex (bitter rot), *Phytophthora* spp., *Alternaria* spp., *Fusarium* spp. and *Stemphylium vesicarium* (Köhl et al., 2018). This considerable diversity of potential pathogens presents challenges to the development of pathogen-specific BCA. Furthermore, the long infection periods over which *Neonectria ditissima* and other pathogens can infect fruits

© Burleigh Dodds Science Publishing Limited, 2019. All rights reserved.

Table 2 Antagonists species successfully used for biological control of fruit rot diseases

Disease (pathogen)	Antagonist	Fruit tree crop
Grey mould (*Botrytis cinerea*)	*Acremonium brevae*	Apple
	Bacillus subtilis	Cherry
	Bacillus pumilus	Pear
	Candida guilliermondii	Nectarine, peach
	Candida sake	Apple
	Cryptococcus laurentii	Peach, pear
	Cryptococcus albidus	Apple
	Kloeckera apiculata	Cherry
	Metschnikowia pulcherrima	Apple
	Pichia guilliermondii	Apple
	Pseudomonas cepacia	Pear
	Pseudomonas fluorescens	Apple
	Pseudomonas syringae	Apple
	Rahuella aquatilis	Apple
	Rhodotorula glutinis	Apple, pear
	Trichoderma harzianum	Pear
Penicillium rot (*Penicillium digitatum, P. expansum*)	*Candida oleophila*	Apple
	Candida sake	Apple, pear
	Cryptococcus laurentii	Peach, pear
	Cryptococcus albidus	Apple
	Cryptococcus sp.	Apple
	Metschnikowia pulcherrima	Apple
	Pantoea agglomerans	Apple
	Pichia guilliermondii	Apple
	Pseudomonas cepacia	Apple, pear
	Pseudomonas syringae	Apple
	Rahuella aquatilis	Apple
	Rhodotorula glutinis	Apple, pear
Monilinia rot (*Monilinia laxa, M. fructicola*)	*Cryptococcus laurentii*	Cherry, peach
	Penicillium frequentans	Peach
	Pseudomonas cepacia	Peach
	Pseudomonas corrugata	Peach, nectarine
	Pseudomonas syringae	Peach
Rhizopus rot (*Rhizopus nigricans, R. stolonifer*)	*Candida sake*	Apple
	Cryptococcus laurentii	Peach
	Debaryomyces hansenii	Peach
	Enterobacter cloacae	Peach
	Pantoea agglomerans	Pear

© Burleigh Dodds Science Publishing Limited, 2019. All rights reserved.

Disease (pathogen)	Antagonist	Fruit tree crop
Mucor rot (*Mucor piriformis*)	*Cryptococcus laurentii*	Pear
	Cryptococcus flavus	Pear
	Cryptococcus albidus	Pear
	Pseudomonas cepacia	Apple
Bitter rot (*Glomerella cingulata*)	*Cryptococcus laurentii*	Apple
Alternaria rot (*Alternaria alternata*)	*Enterobacter aerogenes*	Cherry
Anthracnose (*Colletotrichum gloeosporioides*)	*Pestalotiopsis neglecta*	Apricot

Source: Modified from Sharma et al. (2009).

may require the continuous protection of blossoms and developing fruits by antagonists, which is not easy to achieve. There is a lack of knowledge relating to the epidemiology of this group of post-harvest pathogens so that critical infection periods cannot yet be predicted, preventing efficacious timing of antagonist application.

Generally, fruit rot pathogens have a saprophytic stage which overwinters on necrotic host tissues or necrotic plant tissues of non-hosts present in the orchard. In spring or during the entire growing season such pathogen-colonized necrotic tissues are major sources of pathogen spores which may infect blossoms, developing fruits or mature fruits. Relative importance of different pathogen sources may differ between cropping systems and, more importantly, between different pathogens. For many of the currently increasing number of storage pathogens, epidemiological knowledge regarding their saprophytic stage, such as the relative importance of different necrotic substrates in orchards and conditions for spore production and spore dispersal, is still unknown. Fruit rot pathogens are often found in mummies and fallen leaves (Köhl et al., 2018). Applications of strong saprophytic competitors selected to outcompete a broad range of fruit pathogens on these substrates may be a powerful option to reduce the general risk of fruit rot in orchards.

3.2.1 *Erwinia amylovora* - Fireblight of pome fruit

Fireblight of pome fruit is caused by *Erwinia amylovora*. The main infection sites are flowers, where the pathogen epiphytically colonizes the surfaces of nutrient-rich stigmata (Thomson, 1986; Pusey and Curry, 2004). Under favourable environmental conditions, the pathogen can multiply, move to hypanthia during rainfalls and infect the blossom if present in sufficiently high densities. This particular part of the pathogen life cycle favours the development of antagonists that block the colonization and multiplication of the pathogen on specific flower tissues. The knowledge that flowers represent a specific

© Burleigh Dodds Science Publishing Limited, 2019. All rights reserved.

target, that the flowers open at a specific application time and the timing of the vulnerable pathogen stage - multiplication of the pathogen stimulated by the nutrients available on the stigmata surface - allows a very focussed selection of suitable antagonists. Main ecological characteristics of antagonists are the ability to colonize relevant flower tissues rapidly and to outcompete pathogen populations by competition for nutrients and space, possibly supported by the production of antimicrobial secondary metabolites. Changing the conditions in which bacteria grow in the niche, for example by lowering pH, can be another potential mechanism.

Although the important role of flower infections in the life cycle of *E. amylovora* favours the targeted development of BCA, disease control in the orchard can be limited because individual flowers open at different times so that multiple antagonist applications may be needed. Furthermore, infections of host tissues other than the flowers by the pathogen can occur, so that flower protection will not result in complete disease control.

In recent decades, many antagonists of *E. amylovora* have been described (Johnson and Stockwell, 2000) and a number of these have been developed into commercial biocontrol products (Venneste, 2011). Registered products include those based on bacterial isolates of *Bacillus pumilus, B. subtilis, Pantoea agglomerans, Pseudomonas fluorescens* and the yeast-like fungus *Aureobasidium pullulans* (Table 1). Research on the use of bacteriophages has also showed promising results (Gill et al., 2003; Born et al., 2017) and may open opportunities for future development of additional complimentary products with distinct modes of action.

The scientific successes in the biocontrol of fireblight have not adequately been translated to commercial successes (Venneste, 2011). However, the review of biological agents in the United States in 2003 by Venneste (2011) indicates that there is a substantial use of such agents in addition to the use of the antibiotics streptomycin and oxytetracycline. Venneste (2011) described variability in efficacy of control and difficulties in registration of microorganisms as the two main reasons for the limited market penetration of various biocontrol products. This situation may drastically change in the near future if the use of antibiotics and copper-based pesticides are banned or more strongly regulated.

3.2.2 *Stemphylium vesicarium* - Brown spot of pear

Strains of *Erwinia herbicola* and *Pseudomonas fluorescens* were effective in reducing spore germination and mycelial growth of *Stemphylium vesicarium* in assays under controlled conditions on detached pear leaves and on leaves attached to trees grown under controlled conditions with continuously moist conditions at 22°C (Montesinos et al., 1996). The sensitivity of the bacterial antagonists to low humidity conditions is expected to be the critical factor for

© Burleigh Dodds Science Publishing Limited, 2019. All rights reserved.

their success under field conditions. Alternatively, antagonists were applied to fallen pear leaves on the orchard floor which are a main substrate for *S. vesicarium* for overwintering and for spore production in spring and summer. *Trichoderma* spp. decreased overwintering inoculum of the pathogen (Llorente et al., 2006). Sanitation by leaf removal had only a partial effect but leaf removal combined with application of *Trichoderma* on the orchard floor reduced brown spot incidence by more than 60% in a series of orchard trials in Spain and Italy (Llorente et al., 2010).

3.2.3 *Neonectria ditissima* - European canker in apple

There are few reports on the biocontrol of *Neonectria ditissima*. An isolate of *Bacillus subtilis* applied to defoliated apple trees under field conditions did not reduce canker development (Swinburne, 1973). Walter et al. (2017a) tested the effect of eight BCA containing different strains of *Bacillus subtilis* and *B. subtilis* var. *amyloliquefaciens* by painting canker lesions with highly concentrated products. Sporulation of the pathogen in lesions was reduced under orchard conditions by 13–84%. Canker lesion healing was observed in no case. In another set of orchard trials, freshly made picking wounds and leaf and bud scars were treated with biocontrol products and then contaminated with conidia of *N. ditissima* after 24 h (Walter et al., 2017b). None of the tested products provided wound protection and thus use of biocontrol products during harvest and leaf fall could not be recommended. Schiewe and Mendgen (1992) screened antagonists for post-harvest protection of apple fruit from wound infections by *N. ditissima*. Amongst the more than 700 isolates originating from apple trees and fruit, only four bacterial and three fungal isolates reduced lesions on fruits incubated at 20°C, although at 4°C none of the isolates were effective.

3.3 Other fruit tree diseases

For *Gymnosporangium sabinae* causing European pear rust, no scientific reports on biological control have yet been published. *Bacillus licheniformis* strain SB3086 has registration for *Gymnosporangium* spp. control in the United States (Table 1). Preliminary results on biocontrol of the *Podosphaera leucotricha* causing powdery mildew on apples have been reported by Bosshard et al. (1987). Applications of spore suspensions and culture filtrates of the phyllosphere fungi *Chaetomium* spp. reduced mildew infections on apple seedlings to some extent. It was also found that primary infections of *P. leucotricha* were severely parasitized by *Ampelomyces quisqualis*. However, the spread of the hyperparasite was delayed so that no effect on the secondary powdery mildew epidemic could be observed. *Ampelomyces quisqualis,* effective against various species of *Erysiphales*, has been commercialized under

© Burleigh Dodds Science Publishing Limited, 2019. All rights reserved.

the trade name AQ10 for powdery mildew control in grapevine (Kiss et al., 2004). There are no scientific reports on the use of this product against *P. leucotricha* in apples. Products containing *Bacillus pumilus* QST 2808 and *Bacillus subtilis* QST713 have registration for use against *P. leucotricha* in the United States.

4 Use of biocontrol in orchards

Effective BCA have been selected for control of the main fruit tree diseases. Commercial companies have invested in the product development, registration and marketing of a total of 15 biocontrol products for disease control on fruit tree crops in Australia, Brazil, Canada, European Union, Japan, New Zealand and the United States (Table 1); the majority of these products are registered for control of fireblight or fruit rot. For control of other major diseases such as *Monilinia* and apple scab, antagonists with proven reliable efficacies in orchard trials are ready for commercial product development and some new potential products are already under development and in the early registration stage.

No public data are available on the commercial use of biocontrol products in fruit production. The market share of biocontrol products on the global market for crop protection products is currently around 3.5%, with an annual growth >15% (Glare et al., 2012; van Lenteren et al., 2018). This includes use of macrobial and microbial BCA in various crops. Since the commercial use of biocontrol is most developed in protected crops, it can be estimated that use in fruit tree production is far below 3.5% of the total market for crop protection products.

Although potential biocontrol solutions are available, fruit tree growers use only few biocontrol products and biocontrol manufactures may hesitate to invest in product development and the registration of new products for the fruit tree market. The main reasons for this are (1) the availability of cheap, effective and broad-spectrum chemical disease control products with existing distribution channels, (2) growers are well trained in the use of chemical control products and (3) advisers are highly experienced in the design of complex spray schedules combining various chemical control products, often supported by decision support systems which are developed for optimum timing of chemical control products.

The growth rate of the biocontrol market during the last decade indicates that there is a dynamic shift from chemical control to biocontrol options. Main drivers for this shift are increasing demands for sustainable fruit production systems with low environmental risks, low residues of crop protection products on fruit and the decreasing number of chemical pesticides. The number of registered synthetic chemical crop protection products has decreased significantly in recent years (Glare et al., 2012). The costs for developing new chemical active ingredients have increased significantly and the opportunities

© Burleigh Dodds Science Publishing Limited, 2019. All rights reserved.

for finding new leads have decreased. Furthermore, pathogens continuously develop resistance against currently used fungicides so that their commercial life span is limited.

A combination of efforts is needed to implement biocontrol options in integrated tree fruit productions systems. Communication on the specific strengths and weaknesses of biocontrol solutions must be intensified between researchers, the biocontrol industry, advisers and growers. For example, efficacies of biocontrol products based on living microbial organisms depend much more on environmental conditions compared to chemical control products. Ignoring these dependencies on environmental factors, for example temperature and humidity, will often lead to failure. Thus, a better understanding of antagonist ecology and training of growers, advisers and distributors in the use of biocontrol products will allow the full exploitation of their potential. Generally, unlike most chemical pesticides, biocontrol products are pathogen specific and not broad spectrum. Recent experiences in pome fruit production show an increasing impact of 'minor' fruit rot pathogens and of fungi causing superficial cosmetic disorders (Weber and Zabel, 2011; Wenneker et al., 2017a,b). This may partly be the consequence of the disturbance of fruit microbiomes due to the extensive use of synthetic fungicides with unpredictable side effects on bacterial and fungal communities. On the other hand, side effects of pathogen-specific antagonists on beneficial naturally occurring organisms are not likely, so that natural disease suppression complements biocontrol whereas chemical control often reduces natural suppression, with the risk that new 'iatrogenic' diseases occur in fruit tree crops. Combined multi-year efforts will be needed to develop new integrated application schedules for different biocontrol products for pest and disease control, including the use of modelling (Jeger et al., 2009) and the specific adaptation of decision support systems to the use of biologicals (González-Domínguez et al., 2018).

Essential for this process of broad implementation of biocontrol into commercial fruit tree production systems will be the regulatory environment. Phasing out of groups of crop protection products such as copper-based fungicides for apple scab control and antibiotics for fireblight will have a significant impact. Changes in the current registration procedures are needed to recognize the generally lower and very different nature of the risks associated with biological solutions. Finally, societies may decide to estimate the true costs of plant protection solutions, including the risk of exposure of users and consumers to residues, reduced biodiversity, negative effects on ecosystem services and pollination, and the cleaning of drinking water (Pimentel and Burgess, 2014; Bourguet and Guillemaud, 2016; van Lenteren et al., 2018). These hidden costs of using synthetic pesticides, currently paid externally by societies and not by the manufacturers of the crop protection products, may strongly support the implementation of biocontrol solutions.

© Burleigh Dodds Science Publishing Limited, 2019. All rights reserved.

5 Future trends

The use of microbial biocontrol products in fruit production will increase, as it will for vegetable and arable crops. Main drivers for the increased use of biocontrol products include the loss of available fungicides, development and spread of fungicide resistance of pathogens, environmental considerations and growing demands by retailers and consumers for reduced use of pesticides. There is a significant challenge in moving from the use of biocontrol solutions against a single fruit crop disease to integrated biological programmes with a variety of biocontrol options against major diseases in fruit crops without the need for fungicide interventions. Such a complete set of biocontrol options, including biocontrol of pests, may strengthen the position of each single product.

It is expected that the markets for biocontrol industries will grow with a compound annual growth rate of 16% (Glare et al., 2012). Industrial expertise and infrastructure for mass production of new BCA, and their registration and marketing, will expand accordingly. The underpinning fundamental through to early development research will often be conducted under university and research organizations whereas product development will take place under public-private partnerships. For future research on the development of new BCA two trends are expected to support fruit production. Research into new BCA has often focussed on screening for efficacy; however, a more balanced approach is needed for a broad range of aspects from mass production to ecological characteristics (Glare et al., 2012; Köhl et al., 2011). For example, screening for rhizosphere and phyllosphere competence will result in sufficient persistence on the targets. Microbial inoculants beneficial for fruit crops will be more likely to be studied and exploited with BCA, biostimulants and biofertilizers. Biostimulants contain microorganisms which stimulate natural processes to enhance nutrient uptake, tolerance to stress and crop quality. They can be applied as soil amendment and foliar sprays. Biofertilizers applied to seedlings or soil contain microorganisms which improve phosphate or potassium mobilization or nitrogen fixation. These different objectives for using microbial inoculants are currently being investigated by different research groups, although their practical use is determined by different regulations. However, many of the beneficial microorganisms combine different properties. The same principles of microbial ecology, similar technologies for production, formulation and application, similar manufacturing processes and distribution channels as well as similar principles for assessment for human and environmental risks have to be considered when developing biocontrols, biostimulants and biofertilizers. Overcoming artificial separation of microbial inoculants by their intended use will lead to synergies during product development, marketing and practical use.

Current biocontrol products usually contain one selected bacterial or fungal strain with superior antagonism. Modes of action of antagonists

© Burleigh Dodds Science Publishing Limited, 2019. All rights reserved.

include hyperparasitism, competition for nutrients or space, production of antimicrobial compounds or priming and ISR. Modern 'omics' approaches provide tools to better understand the modes of action of antagonists (Massart et al., 2015). For example, using genomic, transcriptomic, proteomic and metabolomic approaches showed that the metabolism of *Trichoderma* spp., one of the most studied fungal antagonists, is incredibly complex (Lorito et al., 2010). Understanding the genome of *Trichoderma* spp., its functions as a plant symbiont and its interactions with plants and environments is expected to lead to increased product development and better product deployment.

Plants do not rely on a single symbiont but support a whole range of beneficial microorganisms on and in their roots, shoots and leaves (Berg, 2009). Members of the plant microbiome families have complex interactions and usually plants benefit from these interactions. It is expected that exploiting this knowledge on the role of microbiome in biocontrol will lead to the design of synthetic microbial communities (Droby and Wisniewski, 2018; Mazzola and Freilich, 2017). Such synthetic consortia will provide better efficiency, increased robustness and resilience to environmental stress (Droby et al., 2016). Emerging strategies for the use of beneficial microorganisms in crop production will combine microbiome engineering to optimize plant microbiomes and plant breeding to optimize plants in supporting microbiomes with the application of microbial inoculants (Syed Ab Rahman et al., 2018). This will lead to robust exploitation of beneficial microorganisms supporting plant health, nutrient uptake and resilience to environmental stresses.

Knowledge of microbial ecology and the role of microbiomes in crop production are rapidly becoming available due to the implementation of 'omics' approaches. However, the possible use of synthetic consortia in biocontrol is hindered by regulations which were not developed for risk assessments of complex consortia but rather single strains. Current regulation considers members of consortia as a mix of 'active ingredients' which all have to be assessed individually and neglects basic principles of microbial ecology. The key future trends in biocontrol, combing beneficials to form synthetic consortia, needs thus a re-evaluation by the regulatory bodies before these new robust and sustainable solutions will become available for fruit growers.

6 Where to look for further information

Broad scientific information on the principles, ecology and application of biocontrol can be found in the book of Heimpel and Mills (2017), more specific information on the nature of biocontrol of plant pathogens in Cook and Baker (1983) and on the status and success of biocontrol in Nicot (2011). There are recent review articles covering specific aspects of biocontrol of soil pathogens (Berg, 2009; Lugtenberg et al., 2013), leaf pathogens (Elad, 2003; Vorholt,

© Burleigh Dodds Science Publishing Limited, 2019. All rights reserved.

2012), post-harvest pathogens (Droby et al., 2009; Liu et al., 2013; Teixido et al., 2011), mode of action (Massart et al., 2015) and the use of registered BCA (van Lenteren et al., 2018). Results of scientific research on biocontrol of plant pathogens are mainly published in three international journals specifically focussing on biocontrol (*BioControl*, *Biocontrol Science and Technology*, *Biological Control*), with an increasing portion of articles on plant pathogens. However, a significant number of results on biocontrol of plant pathogens are published in a broad range of scientific journals on plant pathology, microbial ecology, soil microbiology, crop-specific journals and journals on post-harvest technologies. The International Organisation for Biological and Integrated Control (IOBC) plays an important role in the communication of scientific results and their integrated use in cropping systems. The IOBC is organized globally with different regional sections. Within the West Palaearctic Regional Section of IOBC (IOBC-WPRS) several working groups include topics on the biocontrol of plant diseases in fruit production: 'IOBC-WPRS WG Integrated protection of fruit crops' with subgroups on 'Pome fruit diseases', 'Stone fruits' and 'Soft fruits', the IOBC-WPRS WG 'Biological and integrated control of plant pathogens', and the IOBC-WPRS WG 'Induced resistance in plants against insects and diseases'. These working groups meet on a regular basis, usually bi-yearly, with international participation of non-members as well. Contributions to the workshops are published in the *IOBC-WPRS Bulletin*. These bulletins usually also include reports on ongoing research and technical papers relating to the practical use of biocontrol in cropping systems.

Information on commercial development, registration and marketing of biocontrol products can be found on the website of the International Biocontrol Manufacturers Association (IBMA; www.ibma-global.org). IBMA organizes annually a global biocontrol industry meeting to network, discover and unveil new products, market opportunities and research areas and inform policy makers and regulators of the specific needs of the biocontrol industry. The website (www.abim.ch) includes an informative archive of presentations held during the meetings since 2006.

There is no central information system on globally available biocontrol products. Products are usually registered by single countries so that national databases on registered crop protection products contain information on biocontrol products including targeted crops and diseases (van Lenteren et al., 2018).

7 References

Andrews, J. H., Berbee, F. M. and Nordheim, E. V. (1983), 'Microbial antagonism to the imperfect stage of the apple scab pathogen, *Venturia inaequalis*', *Phytopathology*, 73, 228–34.

© Burleigh Dodds Science Publishing Limited, 2019. All rights reserved.

Beresford, R. M., Wright, P. J., Wood, P. N., Park, N. M., Larsen, N. J. and Fisher, B. M. (2013), 'Resistance of *Venturia inaequalis* to demethylation inhibitor and dodine fungicides in four New Zealand apple-growing regions', *New Zealand Plant Protection*, 66, 274–83.

Berg, G. (2009), 'Plant-microbe interactions promoting plant growth and health: Perspectives for controlled use of microorganisms in agriculture', *Applied Microbiology and Biotechnology*, 84 (1), 11–18.

Born, Y., Fieseler, L., Thöny, V., Leimer, N., Duffy, B. and Loessner, M. J. (2017), 'Engineering of bacteriophages Y2::dpoL1-C and Y2::luxAB for efficient control and rapid detection of the fire blight pathogen, *Erwinia amylovora*', *Applied and Environmental Microbiology*, 83 (12), article no. e00341-17.

Bosshard, E., Schüpp, H. and Siegfried, W. (1987), 'Concepts and methods in biological control of diseases in apple orchards', *EPPO Bulletin*, 17 (4), 655–63.

Boudreau, M. A. and Andrews, J. H. (1987), 'Factors influencing antagonism of *Chaetomium globosum* to *Venturia inaequalis*: A case study in failed biocontrol', *Phytopathology*, 77, 1470–5.

Bourguet, D. and Guillemaud, T. (2016), The hidden and external costs of pesticide use, *in* Lichtfouse, E. (Ed.), *Sustainable Agriculture Reviews*. Springer, Dordrecht, pp. 35–120.

Burr, T. J., Matteson, M. C., Smith, C. A., Corral-Garcia, M. R. and Huang, T.-C. (1996), 'Effectiveness of bacteria and yeasts from apple orchards as biological control agents of apple scab', *Biological Control*, 6, 151–7.

Carisse, O. (2000), '50 years of research on biological control', *IOBC-WPRS Bulletin*, 23, 5–10.

Carisse, O. and Dewdney, M. (2002), 'A review of non-fungicidal approaches for the control of apple scab', *Phytoprotection*, 83, 1–29.

Carisse, O., Philion, V., Rolland, D. and Bernier, J. (2000), 'Effect of fall application of fungal antagonists on spring ascospore production of apple scab pathogen, *Venturia inaequalis*', *Phytopathology*, 90, 31–7.

Casals, C., Teixido, N., Vinas, I., Silvera, E., Lamarca, N. and Usall, J. (2010), 'Combination of hot water, *Bacillus subtilis* CPA-8 and sodium bicarbonate treatments to control postharvest brown rot on peaches and nectarines', *European Journal of Plant Pathology*, 128, 51–63.

Cook, R. J. and Baker, K. F. (1983), *The Nature and Practice of Biological Control of Plant Pathogens*, American Phytopathological Society.

Droby, S. and Wisniewski, M. (2018), 'The fruit microbiome: A new frontier for postharvest biocontrol and postharvest biology', *Postharvest Biology and Technology*, 140, 107–12.

Droby, S., Wisniewski, M., Macarisin, D. and Wilson, W. (2009), 'Twenty years of postharvest biocontrol research: Is it time for a new paradigm?', *Postharvest Biology and Technology*, 52, 137–45.

Droby, S., Wisniewski, M., Teixidó, N., Spadaro, D. and Jijakli, M. H. (2016), 'The science, development, and commercialization of postharvest biocontrol products', *Postharvest Biology and Technology*, 122, 22–9.

Eilenberg, J., Hajek, A. and Lomer, C. (2001), 'Suggestions for unifying the terminology in biological control', BioControl, 46, 387–400.

Elad, Y. (2003), 'Biocontrol of foliar pathogens: Mechanisms and application', *Communications in Agricultural and Applied Biological Sciences Ghent University*, 68, 17–24.

© Burleigh Dodds Science Publishing Limited, 2019. All rights reserved.

Fiss, M., Kucheryava, N., Schönherr, J., Kollar, A., Arnold, G. and Auling, G. (2000), 'Isolation and characterization of epiphytic fungi from the phyllosphere of apple as potential biocontrol agents against apple scab (*Venturia inaequalis*)', *Zeitschrift für Pflanzenkrankheiten und Pflanzenschutz / Journal of Plant Diseases and Protection*, 107, 1-11.

Fokkema, N. J. and de Nooij, M. P. (1981), 'The effect of fungicides on the microbial balance in the phyllosphere', *EPPO Bulletin*, 11(3), 303-10.

Gill, J. J., Svircev, A. M., Smith, R. and Castle, A. J. (2003), 'Bacteriophages of *Erwinia amylovora*', *Applied and Environmental Microbiology*, 69 (4), 2133-8.

Glare, T., Caradus, J., Gelernter, W., Jackson, T., Keyhani, N., Köhl, J., Marrone, P., Morin, L. and Stewart, A. (2012), 'Have biopesticides come of age?', *Trends in Biotechnology*, 30, 250-8.

Gomez, C., Brun, L., Chauffour, D. and Vallée, D. D. L. (2007), 'Effect of leaf litter management on scab development in an organic apple orchard', *Agriculture, Ecosystems & Environment*, 118, 249-55.

González-Domínguez, E., Legler, S., Fedele, G., Si Ammour, M., Caffi, T. and Rossi, V. (2018), 'Helping farmers in timing the application of biocontrol agents in viticulture', *IOBC-WPRS Bulletin*, 133, 73-5.

Gotor-Vila, A., Teixidó, N., Usall, J., Dashevskaya, S. and Torres, R. (2016), 'Development of a SCAR marker and a strain-specific genomic marker for the detection of the biocontrol agent strain CPA-8 *Bacillus amyloliquefaciens* (formerly *B. subtilis*)', *Annals of Applied Biology*, 169 (2), 248-56.

Gotor-Vila, A., Teixidó, N., Casals, C., Torres, R., De Cal, A., Guijarro, B. and Usall, J. (2017a), 'Biological control of brown rot in stone fruit using *Bacillus amyloliquefaciens* CPA-8 under field conditions', *Crop Protection*, 102, 72-80.

Gotor-Vila, A., Usall, J., Torres, R., Abadias, M. and Teixidó, N. (2017b), 'Formulation of the biocontrol agent *Bacillus amyloliquefaciens* CPA-8 using different approaches: Liquid, freeze-drying and fluid-bed spray-drying', *BioControl*, 62 (4), 545-55.

Grzegorczyk, M., Żarowska, B., Restuccia, C. and Cirvilleri, G. (2017), 'Postharvest biocontrol ability of killer yeasts against *Monilinia fructigena* and *Monilinia fructicola* on stone fruit', *Food Microbiology*, 61, 93-101.

Guijarro, B., Larena, I., Melgarejo, P. and De Cal, A. (2017), 'Adaptive conditions and safety of the application of *Penicillium frequentans* as a biocontrol agent on stone fruit', *International Journal of Food Microbiology*, 254, 25-35.

Heimpel, G. E. and Mills, N. (2017), *Biological Control - Ecology and Applications*, Cambridge University Press.

Heye, C. C. and Andrews, J. H. (1983), 'Antagonism of *Athelia bombacina* and *Chaetomium globosum* to the apple scab pathogen *Venturia inaequalis*', *Phytopathology*, 73, 650-4.

Janisiewicz, W. J., Kurtzman, C. P. and Buyer, J. S. (2010), 'Yeasts associated with nectarines and their potential for biological control of brown rot', *Yeast*, 27, 389-98.

Janisiewicz, W. J., Jurick, W. M., Peter, K. A., Kurtzman, C. P. and Buyer, J. S. (2014), 'Yeasts associated with plums and their potential for controlling brown rot after harvest', *Yeast*, 31 (6), 207-18.

Jeger, M. J., Jeffries, P., Elad, Y. and Xu, X.-M. (2009), 'A generic theoretical model for biological control of foliar plant diseases', *Journal of Theoretical Biology*, 256 (2), 201-14.

© Burleigh Dodds Science Publishing Limited, 2019. All rights reserved.

Johnson, K. B. and Stockwell, V. O. (2000), Biological control of fire blight, *in* Vanneste, J. L. (Ed.), *Fire Blight: The Disease and Its Causative Agent, Erwinia amylovora*. CAB International, Wallingford, UK, pp. 319-37.

Kiss, L., Russell, J. C., Szentiványi, O., Xu, X. and Jeffries, P. (2004), 'Biology and biocontrol potential of Ampelomyces mycoparasites, natural antagonists of powdery mildew fungi', *Biocontrol Science and Technology*, 14, 635-51.

Köhl, J. A. (2009),'Novel micro-organisms controlling plant pathogens', *International Patent Application WO 2009/078710*.

Köhl, J. and Fokkema, N. J. (1998), 'Biological control of necrotrophic foliar fungal pathogens', *in* Boland, G. J. and Kuykendall, L. V. (Eds), *Plant-Microbe Interactions and Biological control*. Marcel Dekker, New York, pp. 49-88.

Köhl, J., Molhoek, W. M. L., Groenenboom-de Haas, B. H. and Goossen-van de Geijn, H. M. (2009), 'Selection and orchard testing of antagonists suppressing conidia production of the apple scab pathogen *Venturia inaequalis*', *European Journal of Plant Pathology*, 123, 401-14.

Köhl, J., Postma, J., Nicot, P., Ruocco, M. and Blum, B. (2011), 'Stepwise screening of microorganisms for commercial use in biological control of plant pathogenic fungi and bacteria', *Biological Control*, 57, 1-12.

Köhl, J., Scheer, C., Holb, I. J., Masny, S. and Molhoek W. M. L. (2015), 'Towards an integrated use of biological control by *Cladosporium cladosporioides* H39 in apple scab (*Venturia inaequalis*) management', *Plant Disease*, 99, 535-43.

Köhl, J., Wenneker, M., Groenenboom-de Haas, B. H., Anbergen, R., Goossen-van de Geijn, H. M., Lombaers-van der Plas, C. H., Pinto, F. A. M. F. and Kastelein, P. (2018), 'Dynamics of post-harvest pathogens *Neofabraea* spp. and *Cadophora* spp. in plant residues in Dutch apple and pear orchards', *Plant Pathology*. doi:10.1111/ppa.12854.

Larena, I., Torres, R., De Cal, A., Liñán, M., Melgarejo, P., Domenichini, P., Bellini, A., Mandrin, J. F., Lichou, J., De Eribe, X. O. and Usall, J. (2005), 'Biological control of postharvest brown rot (*Monilinia* spp.) of peaches by field applications of *Epicoccum nigrum*', *Biological Control*, 32 (2), 305-10.

Liu, J., Sui, Y., Wisniewski, M., Droby, S. and Liu, Y. (2013), 'Review: Utilization of antagonistic yeasts to manage postharvest fungal diseases of fruit', *International Journal of Food Microbiology*, 167 (2), 153-60.

Llorente, I., Vilardell, A. and Montesinos, E. (2006), 'Infection potential of *Pleospora allii* and evaluation of methods for reduction of the overwintering inoculum of brown spot of pear', *Plant Disease*, 90 (12), 1511-16.

Llorente, I., Vilardell, A., Vilardell, P., Pattori, E., Bugiani, R., Rossi, V. and Montesinos, E. (2010), 'Control of brown spot of pear by reducing the overwintering inoculum through sanitation', *European Journal of Plant Pathology*, 128 (1), 127-41.

Lorito, M., Woo, S. L., Harman, G. E. and Monte, E. (2010), 'Translational research on trichoderma: From 'omics to the field', *Annual Review of Phytopathology*, 48, 395-417.

Lugtenberg, B. J. J., Malfanova, N., Kamilova, F. and Berg, G. (2013), 'Microbial control of plant root diseases', *Molecular Microbial Ecology of the Rhizosphere*, 2, 575-86.

MacHardy, W. E., Gadoury, D. M. and Gessler, C. (2001), 'Parasitic and biological fitness of *Venturia inaequalis*: Relationship to disease management strategies', *Plant Disease*, 85, 1036-51.

© Burleigh Dodds Science Publishing Limited, 2019. All rights reserved.

Massart, S., Perazzolli, M., Höfte, M., Pertot, I. and Jijakli, M. H. (2015), 'Impact of the omic technologies for understanding the modes of action of biological control agents against plant pathogens', *BioControl*, 60 (6), 725–46.

Mazzola, M. and Freilich, S. (2017), 'Prospects for biological soilborne disease control: Application of indigenous versus synthetic microbiomes', *Phytopathology*, 107, 256–63.

Melgarejo, P., Carrillo, R. and M-Sagasta, E. (1985), 'Mycofora of peach twigs and flowers and its possible significance in biological control of *Monilinia laxa*', *Transactions of the British Mycological Society*, 85, 313–17.

Melgarejo, P., Carrillo, R. and M-Sagasta, E. (1986), 'Potential for biological control of *Monilinia laxa* in peach twigs', *Crop Protection*, 5, 422-6.

Miedtke, U. and Kennel, W. (1990), '*Athelia bombacina* and *Chaetomium globosum* as antagonists of the perfect stage of the apple scab pathogen (*Venturia inaequalis*) under field conditions', *Zeitschrift für Pflanzenkrankheiten und Pflanzenschutz / Journal of Plant Diseases and Protection*, 97, 24–32.

Montesinos, E., Bonaterra, A., Ophir, Y. and Beer, S. V. (1996), 'Antagonism of selected bacterial strains to *Stemphylium vesicarium* and biological control of brown spot of pear under controlled environment conditions', *Phytopathology*, 86 (8), 856–63.

Nicot, P. C. (2011), *Classical and Augmentative Biological Control against Diseases and Pests: Critical Status Analysis and Review of Factors Influencing Their Success*, IOBC-WPRS.

Ortega, F., Steiner, U. and Dehne, H.-W. (1998), 'Induced resistance to apple scab: Microscopic studies on the infection cycle of *Venturia inaequalis* (Cke.) Wint.', *Journal of Phytopathology*, 146, 399–405.

Percival, G. C., Noviss, K. and Haynes, I. (2009), 'Field evaluation of systemic inducing resistance chemicals at different growth stages for the control of apple (*Venturia inaequalis*) and pear (*Venturia pirina*) scab', *Crop Protection*, 28, 629–33.

Pimentel, D. and Burgess, M. (2014), 'Environmental and economic costs of the application of pesticides primarily in the United States', *in* Pimentel, D. and Peshin, R. (Eds), *Integrated Pest Management*. Springer, Dordrecht, pp. 47-71.

Piombo, E., Sela, N., Wisniewski, M., Hoffmann, M., Gullino, M. L., Allard, M. W., Levin, E., Spadaro, D. and Droby, S. (2018), 'Genome sequence, assembly and characterization of two *Metschnikowia fructicola* strains used as biocontrol agents of postharvest diseases', *Frontiers in Microbiology*, 9, art. no. 593. doi:10.3389/fmicb.2018.00593.

Pusey, P. L. and Curry, E. A. (2004), 'Temperature and pomaceous flower age related to colonization by *Erwinia amylovora* and antagonists', *Phytopathology*, 94, 901–11.

Pusey, P. L. and Wilson, C. L. (1984), 'Postharvest biological control of stone fruit brown rot by *Bacillus subtilis*', *Plant Disease*, 68, 753–6.

Pusey, P. L., Hotchkiss, M. W., Dulmage, H. T., Baumgardner, R. A., Zehr, E. I., Reilly, C. C. and Wilson, C. L. (1988), 'Pilot tests for commercial production and application of *Bacillus subtilis* (B-3) for postharvest control of peach brown rot', *Plant Disease*, 72, 622–6.

Rungjindamai, N., Jeffries, P. and Xu, X.-M. (2014a), 'Epidemiology and management of brown rot on stone fruit caused *by Monilinia laxa*', *European Journal of Plant Pathology*, 140 (1), 1–17.

Rungjindamai, N., Jeffries, P. and Xu, X.-M. (2014b), 'A novel strategy to reduce overwintering inoculum of *Monilinia laxa*', *European Journal of Plant Pathology*, 140 (3), 591–6.

© Burleigh Dodds Science Publishing Limited, 2019. All rights reserved.

Schiewe, A. and Mendgen, K. (1992), 'Identification of antagonists for biological control of the post-harvest pathogens *Pezicula malicorticis* and *Nectria galligena* on apples', *Journal of Phytopathology*, 134 (3), 229–37.

Sharma, R. R., Singh, D. and Singh, R. (2009), 'Biological control of postharvest diseases of fruits and vegetables by microbial antagonists: A review', *Biological Control*, 50 (3), 205–21.

Snowdon, A. L. (1990), 'Pome fruits', *in* Snowdon, A. L. (Ed.), *A Colour Atlas of Post-harvest Diseases and Disorders of Fruits and Vegetables. Vol. 1: General Introduction and Fruits*. Wolfe Scientific Ltd., London, pp. 170–218.

Spadaro, D. and Droby, S. (2016), 'Development of biocontrol products for postharvest diseases of fruit: The importance of elucidating the mechanisms of action of yeast antagonists', *Trends in Food Science and Technology*, 47, 39–49.

Sutton, D. K., MacHardy, W. E. and Lord, W. G. (2000), 'Effects of shredding or treating apple leaf litter with urea on ascospore dose of *Venturia inaequalis* and disease buildup', *Plant Disease*, 84, 1319–26.

Swinburne, T. R. (1973), 'Microflora of apple-leaf scars in relation to infection by *Nectria galligena*', *Transactions of the British Mycological Society*, 60, 389–403.

Syed Ab Rahman, S. F., Singh, E., Pieterse, C. M. J. and Schenk, P. M. (2018), 'Emerging microbial biocontrol strategies for plant pathogens', *Plant Science*, 267, 102–11.

Teixido, N., Torres, R., Viñas, I., Abadias, M. and Usall, J. (2011), 'Biological control of postharvest diseases in fruit and vegetables', *in Lacroix* (Ed.), *Protective Cultures, Antimicrobial Metabolites and Bacteriophages for Food and Beverage Biopreservation*. Woodhead Publishing. doi:10.1533/9780857090522.3.364.

Thomson, S. V. (1986), 'The role of the stigma in fire blight infections', *Phytopathology*, 76, 476–82.

van Lenteren, J. C., Bolckmans, K., Köhl, J., Ravensberg, W. and Urbaneja, A. (2018), 'Biological control using invertebrates and microorganisms: Plenty of new opportunities', *BioControl*, 63, 39–59. doi:10.1007/s10526-017-9801-4.

Venneste, J. L. (2011), 'Biological control agents of fire blight: Successes and challenges', *Acta Horticulturae*, 896, 409–16.

Vorholt, J. A. (2012), 'Microbial life in the phyllosphere', *Nature Reviews Microbiology*, 10 (12), 828–40.

Walter, M., Campbell, R. E., Amponsah, N. T., Turner, L., Rainham, D., Kerer, U. and Butler, R. C. (2017a), 'Can biological products control *Neonectria ditissima* picking wound and leaf scar infections in apples?', *New Zealand Plant Protection*, 70, 63–72.

Walter, M., Campbell, R. E., Amponsah, N. T., Scheper, R. W. A. and Butler, R. C. (2017b), 'Evaluation of biological and agrichemical products for control of *Neonectria ditissima* conidia production', *New Zealand Plant Protection*, 70, 87–96.

Weber, R. W. S. and Zabel, D. (2011), 'White haze and scarf skin, two little-known cosmetic defects of apples in Northern Germany', *European Journal of Horticultural Science*, 76(2), 45–50.

Wenneker, M., Pham K. T. K., Boekhoudt, L. C., de Boer, F. A., van Leeuwen, P. J., Hollinger, T. C. and Thomma, B. P. H. J. (2017a), 'First report of *Neofabraea kienholzii* causing bull's eye rot on pear (*Pyrus communis*) in the Netherlands', *Plant Disease*, 101, 634.

Wenneker, M., Pham, K. T. K., Lemmers, M. E. C., de Boer, F. A., van Leeuwen, P. J., Hollinger, T. C., van de Geijn, F. G. and Thomma, B. P. H. J. (2017b), '*Fibulorhizoctonia psychrophila* is the causal agent of lenticel spot on apple and pear fruit in the

© Burleigh Dodds Science Publishing Limited, 2019. All rights reserved.

Netherlands', *European Journal of Plant Pathology*, 148 (1), 213–17. DOI: 10.1007/s10658-016-1061-0.
Zhang, J., Xie, J., Zhou, Y., Deng, L., Yao, S. and Zeng, K. (2017), 'Inhibitory effect of *Pichia membranaefaciens* and *Kloeckera apiculata* against *Monilinia fructicola* and their biocontrol ability of brown rot in postharvest plum', *Biological Control*, 114, 51–8.
Zhou, T., Schneider, K. E. and Li, X. (2008), 'Development of biocontrol agents from food microbial isolates for controlling post-harvest peach brown rot caused by *Monilinia fructicola*', *International Journal of Food Microbiology*, 126, 180–5.

© Burleigh Dodds Science Publishing Limited, 2019. All rights reserved.

www.ingramcontent.com/pod-product-compliance
Lightning Source LLC
LaVergne TN
LVHW010131311026
851422LV00030B/282
9781835457016